Molecular Forensics

Molecular Forensics

Edited by

Ralph Rapley

Department of Biosciences, University of Hertfordshire, Hatfield, UK

David Whitehouse

Pathogen Molecular Biology Unit, London School of Hygiene and Tropical Medicine, London, UK

John Wiley & Sons, Ltd

Copyright © 2007   John Wiley & Sons Ltd, The Atrium, Southern Gate, Chichester,
West Sussex PO19 8SQ, England

Telephone   (+44) 1243 779777

Email (for orders and customer service enquiries): cs-books@wiley.co.uk
Visit our Home Page on www.wiley.co.uk or www.wiley.com

Reprinted June 2008

Other Wiley Editorial Offices

John Wiley & Sons Inc., 111 River Street, Hoboken, NJ 07030, USA

Jossey-Bass, 989 Market Street, San Francisco, CA 94103-1741, USA

Wiley-VCH Verlag GmbH, Boschstr. 12, D-69469 Weinheim, Germany

John Wiley & Sons Australia Ltd, 33 Park Road, Milton, Queensland 4064, Australia

John Wiley & Sons (Asia) Pte Ltd, 2 Clementi Loop #02-01, Jin Xing Distripark, Singapore 129809

John Wiley & Sons Canada Ltd, 6045 Freemont Blvd, Mississauga, ONT, L5R 4J3, Canada

Wiley also publishes its books in a variety of electronic formats. Some content that appears in print
may not be available in electronic books.

Anniversary Logo Design: Richard J. Pacifico

Library of Congress Cataloging-in-Publication Data

Molecular forensics / edited by Ralph Rapley, David Whitehouse.
    p. ; cm.
  Includes bibliographical references and index.
  ISBN-13: 978-0-470-02495-9 (hb. : alk. paper)
  ISBN-13: 978-0-470-02496-6 (alk. paper)
1. Forensic biology. 2. Chemistry, Forensic. I. Rapley, Ralph. II. Whitehouse, David
  [DNLM: 1. Forensic Medicine – methods. 2. DNA – analysis. 3. Genetic Techniques.
W 700 M718 2007]
  QH313.5.F67M65 2007
  363.25 – dc22   2006038357

British Library Cataloguing in Publication Data

A catalogue record for this book is available from the British Library

ISBN 978-0-470-02495-9 H/B
    978-0-470-02496-6 P/B

Typeset in 10.5 on 12.5 pt Sabon by SNP Best-set Typesetter Ltd., Hong Kong
Printed and bound in Great Britain by CPI Antony Rowe, Chippenham, Wiltshire

Contents

Preface

Just over twenty years ago Alec Jeffreys laid the foundation stone of modern molecular forensics with the discovery of hypervariable minisatellites and DNA fingerprinting. Before that time, precise human individualization was not feasible and the best that could be achieved by forensic scientists was an exclusion probability based on information from gene product analysis of polymorphic blood groups and protein loci or restriction fragment length polymorphisms. In contrast, with the exception of monozygotic twins, DNA fingerprints possessed the capacity to match a sample to a unique individual and this capacity to positively individualize changed the mindset of forensic science forever. Progress in forensic analysis has been rapid since the introduction of the multilocus probes, through single-locus VNTR probes and microsatellite loci to incoming approaches including single nucleotide polymorphisms (SNPs) and proteomic microarrays.

This volume brings together a collection of chapters by internationally recognized authors. The opening chapter by Simon Walsh describes the process of development within the forensic molecular biology field. It also touches briefly on the way such developments intersect with the neighbouring fields of law enforcement and the justice system. The following three chapters are dedicated to current molecular biology approaches. In Chapter 2 the principal molecular techniques are summarized, including DNA extraction, polymerase chain reaction (PCR) and DNA sequencing. The next two chapters are more detailed treatments of important technical issues. In Chapter 3 Marion Nagy examines automated approaches for isolating DNA from biological material and in Chapter 4 Antonio Alonso and Oscar Fernández review the different real-time PCR assays that have been applied in forensic genetics for human analysis along with some real-time PCR assays for non-human species of forensic. The following two chapters are devoted to the main polymorphic systems available to forensic science. In Chapter 5, Keiji Tamaki describes the historical development and current forensic applications of polymorphic minisatellites and microsatellites, currently the preferred choice for forensic DNA analysis. Claus Børsting and colleagues in Chapter 6 provide an insight into the potential forensic applications of SNPs, including the possibility of identifying phenotypic characteristics from SNPs located in coding regions. The following three chapters deal with polymorphic markers located on specific chromosomes: chromosome X is dealt with by Reinhard Szibor, the mitochondrial genome, which is currently under intense investigation, by Hirokazu Matsuda and Nobuhiro

Yukawa and the Y chromosome by Manfred Keyser. The remaining chapters bring together a variety of key issues in forensic science. Laser microdissection, Chapter 10, is reviewed by Luigi Saravo and colleagues. In Chapter 11, Benoît Leclair and Tom Scholl address the crucial issues that arise with the management of forensic laboratory data. Following this, Mark Best presents the statistical background to the analysis of polymorphism data and its use in casework. The book concludes with two forward-looking contributions. Mikhail Soloviev and colleagues evaluate the use of known protein markers for quantitative protein profiling assays applicable to forensic and biometric applications, and Chris Boesch and colleagues assess the application of magnetic resonance spectroscopy to the persistent difficulty of estimating the postmortem interval.

Ralph Rapley and David Whitehouse

List of Contributors

Antonio Alonso
Instituto Nacional de Toxicología y
Ciencias Forenses
Servicio de Biología, Luis Cabrera 9
28002 Madrid, Spain

Julian Bailes
School of Biological Sciences
Royal Holloway, University of London
Egham, Surrey TW20 0EX, UK

Mark A. Best
Lake Erie College of Osteopathic
Medicine
Bradenton, FL, USA

Chris Boesch
Department of Clinical Research
MR-Spectroscopy and Methodology
University of Bern, Switzerland

Claus Børsting
Department of Forensic Genetics
Institute of Forensic Medicine
University of Copenhagen
DK-2100 Copenhagen, Denmark

Ignazio Ciuna
Molecular Biology Division
Raggruppamento Carabinieri
Investigazioni Scientifiche
S.S.114 Km 6,400-98128 Tremestieri
(ME)
Italy

Enrico Di Luise
Molecular Biology Division
Raggruppamento Carabinieri
Investigazioni Scientifiche
S.S.114 Km 6,400-98128 Tremestieri
(ME)
Italy

Davide Di Martino
Molecular Biology Division
Raggruppamento Carabinieri
Investigazioni Scientifiche
S.S.114 Km 6,400-98128 Tremestieri
(ME)
Italy

Paul Finch
School of Biological Sciences
Royal Holloway, University of London
Egham, Surrey TW20 0EX, UK

Oscar García
Área de Laboratorio Ertzaintza
Sección de Genética Forense
Erandio, Bizkaia
Basque Country, Spain

Ernesto Ginestra
Molecular Biology Division
Raggruppamento Carabinieri
Investigazioni Scientifiche
S.S.114 Km 6,400-98128 Tremestieri
(ME)
Italy

Giuseppe Giuffrè
Laboratory of Applied Molecular Biology
Department of Human Pathology
University of Messina, Italy

Michael Ith
Institute of Forensic Medicine
University of Bern, Switzerland

Manfred Kayser
Department of Forensic Molecular
Biology
Erasmus University Medical Centre
Rotterdam
Medical-Genetic Cluster, PO Box 1738
3000 DR Rotterdam, The Netherlands

Benoît Leclair
Myriad Genetic Laboratories, Inc.
320 Wakara Way, Salt Lake City
UT 84108, USA

Beniamino Leo
Molecular Biology Division
Raggruppamento Carabinieri
Investigazioni Scientifiche
S.S.114 Km 6,400-98128 Tremestieri
(ME)
Italy

Hirokazu Matsuda
Division of Legal Medicine
Department of Social Medicine
Faculty of Medicine
University of Miyazaki
5200 Kihara, Kiyotake-cho
Miyazaki 889-1692, Japan

Niels Morling
Department of Forensic Genetics
Institute of Forensic Medicine
University of Copenhagen
DK-2100 Copenhagen, Denmark

Marion Nagy
Institute of Legal Medicine
Charité-University Medicine Berlin
Hannoversche Strasse 6
10115 Berlin, Germany

Dario Piscitello
Molecular Biology Division
Raggruppamento Carabinieri
Investigazioni Scientifiche
S.S.114 Km 6,400-98128 Tremestieri
(ME)
Italy

Fabio Quadrana
Molecular Biology Division
Raggruppamento Carabinieri
Investigazioni Scientifiche
S.S.114 Km 6,400-98128 Tremestieri
(ME)
Italy

Ralph Rapley
Department of Biosciences
University of Hertfordshire
Hatfield AL 10 9AB, UK

Carlo Romano
Molecular Biology Division
Raggruppamento Carabinieri
Investigazioni Scientifiche
S.S.114 Km 6,400-98128 Tremestieri
(ME)
Italy

Nina Salata
School of Biological Sciences
Royal Holloway, University of London
Egham, Surrey TW20 0EX, UK

Juan J. Sanchez
Department of Forensic Genetics
Institute of Forensic Medicine
University of Copenhagen
DK-2100 Copenhagen, Denmark

Luigi Saravo
Molecular Biology Division
Raggruppamento Carabinieri
Investigazioni Scientifiche
S.S.114 Km 6,400-98128 Tremestieri
(ME)
Italy

Eva Scheurer
Institute of Forensic Medicine
University of Bern, Switzerland

Tom Scholl
Genzyme Genetics
3400 Computer Drive
Westborough, MA 01581, USA

Mikhail Soloviev
School of Biological Sciences
Royal Holloway, University of London
Egham, Surrey TW20 0EX, UK

Salvatore Spitaleri
Molecular Biology Division
Raggruppamento Carabinieri
Investigazioni Scientifiche
S.S.114 Km 6,400-98128 Tremestieri
(ME)
Italy

Nicola Staiti
Molecular Biology Division
Raggruppamento Carabinieri
Investigazioni Scientifiche
S.S.114 Km 6,400-98128 Tremestieri
(ME)
Italy

Reinhard Szibor
Institut für Rechtsmedizin
Otto-von-Guericke-Universität
Magdeburg
Leipziger Strasse 44
D-39120 Magdeburg, Germany

Keiji Tamaki
Department of Legal Medicine
Kyoto University Graduate School of
Medicine
Kyoto 606-8501, Japan

Giovanni Tuccari
Laboratory of Applied Molecular Biology
Department of Human Pathology
University of Messina, Italy

Simon J. Walsh
Biological Criminalistics
Forensic and Technical Services
Australian Federal Police
Canberra, ACT, 2611
Australia

David Whitehouse
London School of Hygiene and Tropical
Medicine
Pathogen Molecular Biology Unit
Department of Infectious and Tropical
Diseases
Keppel Street, London WC1E 7HT, UK

Nobuhiro Yukawa
Division of Legal Medicine
Department of Social Medicine
Faculty of Medicine
University of Miyazaki
5200 Kihara, Kiyotake-cho
Miyazaki 889-1692, Japan

1
Current and future trends in forensic molecular biology

Simon J. Walsh

1.1 Introduction

Forensic science is part of a process beginning at a crime scene and concluding in a court room. This means that as one of the key forensic disciplines, the field of forensic molecular biology resides within the complex and adversarial context of the criminal justice system (CJS). The key areas of the CJS that are relevant to the use of forensic molecular biology are the domains of law enforcement and the justice system (Figure 1.1). Due to the intersection of these three domains, changes and developments in one can have a resultant impact on the other adjacent areas. Therefore, when considering the current and future trends in forensic molecular biology it is important to do so not only from the perspective of their effect within the forensic field itself, but also from the perspective of their interaction with neighbouring areas of the system. After all, it is in these neighbouring areas that forensic outcomes are eventually put to use.

Forensic molecular biology has developed rapidly into a comprehensive discipline in its own right and, perhaps more so than any scientific advance before it, has had a profound impact across the CJS. Within the forensic science discipline, as expected, development has been science and/or technology driven. It has followed a trend towards achieving greater sophistication, throughput and informativeness for the DNA-based outcomes of scientific analysis. Developments in forensic molecular biology that have influenced law enforcement could be thought of as operational developments as they predominantly apply to the manner or degree that forensic molecular biology is utilized. As such, they typically have both a technical and policy-oriented basis. Progress in forensic biology

Molecular Forensics. Edited by Ralph Rapley and David Whitehouse
Copyright 2007 by John Wiley & Sons, Ltd.

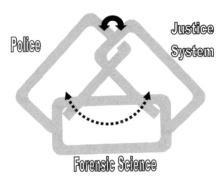

Figure 1.1   A simplified representation of the areas of the CJS that are relevant to forensic molecular biology. A large number of cases flow directly between the police and legal domains (solid arrow) whereas a reduced number of cases flow through the forensic domain (dashed arrow). Each of the three areas can be thought of as intersecting and, as such, each has the capacity to exercise some effect on the others

has also influenced the justice sector. This is characterized, for example, by the iterative response of both the legislature and the courts to changes in the volume and nature of forensic DNA tests. Throughout the history of the field there has also been associated debate and controversy accompanying these legal developments. This reflects the array of socio-legal and ethical issues associated with more widespread use of forensic molecular biology.

This chapter chiefly describes the process of development within the forensic molecular biology field. It also touches briefly on the way such developments intersect with the neighbouring fields of law enforcement and the justice system. By considering developmental trends in this way the overall impact of changes in forensic molecular biology can be appropriately placed in context, allowing reflection on their effect to date and foreshadowing their potential effect in the future.

1.2   Developments within the field of forensic molecular biology

From the time the field settled on a uniform technological platform (Gill, 2002), forensic molecular biologists have done a masterful job at extending the applicability of this testing regime as far as conceivably possible. The discriminating power of short tandem repeat (STR)-based tests has been increased by combining up to 16 (Collins *et al.*, 2000; Krenke *et al.*, 2002) STR loci into a single polymerase chain reaction (PCR; see Chapter 2). The sensitivity of the routine tests has also been driven downward so that successful analysis is now achieved from as little as 100 pg of starting template (Whitaker *et al.*, 2001).

Advancing the capabilities of the DNA methodology has also expanded the range of criminal cases and sample types able to be successfully analysed. For many years forensic molecular biology was limited to testing templates such as blood, semen, hair and saliva. However, the increased efficiency of the STR-based methods now means that DNA can be successfully analysed from discarded clothing or personal effects (Webb *et al.*, 2001), skin cell debris from touched or handled surfaces (Van Oorschot and Jones, 1997; Wiegand and Kleiber, 1997; Zamir *et al.*, 2000; Bright and Petricevic, 2004), dandruff (Lorente *et al.*, 1998), drinking containers (Abaz *et al.*, 2002), food (Sweet and Hildebrand, 1999) and fingernail clippings and scrapings (Harbison *et al.*, 2003). Recent approaches such as reduced-amplicon STR analysis (Butler *et al.*, 1998, 2003; Wiegand and Kleiber, 2001; Coble and Butler, 2005) and low copy number (LCN) profiling (Gill, 2001a; Whitaker *et al.*, 2001) have enhanced reaction sensitivity even further and improved the ability to analyse the most troublesome and highly degraded samples.

Many of the routine techniques have been adapted onto automated platforms so as to facilitate high-throughput analysis and reduce the amount of sample handling (Gill, 2002; Varlaro and Duceman, 2002; Fregeau *et al.*, 2003) (see Chapter 3). Computer-assisted data analysis has also further streamlined the analytical process and reduced some areas of subjectivity, such as mixture interpretation (Perlin and Szabady, 2001; Bill *et al.*, 2005; Gill *et al.*, 2005, 2006b) (see Chapter 11). The next generation of laboratory instrumentation includes micro-scale electrophoresis devices (Woolley *et al.*, 1997; Mitnik *et al.*, 2002) that not only promise rapid analysis times but also allow for the possibility of remote or portable laboratory platforms (Hopwood *et al.*, 2006).

The observable trend in the development areas mentioned above is that they are all directed towards improving the ability to undertake routine DNA-based identity testing. Whilst this refinement of routine typing technologies is of vital importance, it has meant that for the most part the field has sought only one dimension of information from biological evidence samples. Through recent research into the physical and genetic properties of human DNA this is now changing, allowing the forensic field to diversify its capabilities and begin to address questions beyond the identification of source.

There are already several examples of forensic molecular biology applications that either apply different forms of typing technologies or address a different line of genetic inquiry via new polymorphisms or loci. One such area is non-autosomal DNA profiling, particularly the analysis of mitochondrial DNA (mtDNA) and Y chromosome markers (see Chapters 7, 8 and 9). Whilst mtDNA analysis has been widely used in human evolutionary biology for a number of years (Cann *et al.*, 1987), its routine application to forensic work has been consistently evolving. In forensic science, mtDNA is most often analysed in circumstances where nuclear DNA fails to give a result, such as in the analysis of telogenic hairs (Wilson *et al.*, 1995), nail material (Anderson *et al.*, 1999) and bone (Bender *et al.*, 2000; Edson *et al.*, 2004) or when distant relatives must be used as reference (Gill *et al.*, 1994; Ivanov *et al.*, 1996; Pfeiffer *et al.*, 2003). Analysis typically involves direct sequencing of the hypervariable regions 1 and 2 (HV1 and HV2, respectively) (Tully *et al.*, 2001) although SNP-based approaches offer the potential to complement or substitute the need for sequencing (Budowle *et al.*, 2004; Coble *et al.*, 2004; Quintans *et al.*, 2004). Recent developmental progress in the forensic use of mtDNA has also been shaped by the context within which it has been required. In particular, the large-scale multi-national response to recent wars (Huffine *et al.*, 2001; Andelinovic *et al.*, 2005), refugee crises (Lorente *et al.*, 2002) and mass fatalities (Roby, 2002; Vastag, 2002; Budjimila *et al.*, 2003; Holland *et al.*, 2003; Budowle *et al.*, 2005) has seen a rapid evolution of these and other specialist identification sciences so as to respond to the unprecedented logistical and technical challenges presented by these circumstances.

The analysis of polymorphisms on the non-recombining portion of the human Y chromosome (NRY) (Jobling *et al.*, 1997; Kayser *et al.*, 1997) has also steadily developed into a valuable forensic technique (Gill *et al.*, 2001; Gusmao and Carracedo, 2003; Gusmao *et al.*, 2006). The male specificity of the Y chromosome makes it particularly suitable for the resolution of problematic situations such as complex mixtures. In a casework setting Y chromosome analysis is especially useful for typing mixed male–female stains that commonly occur as a result of sexual assaults (Dettlaff-Kakol and Pawlowski, 2002; Dziegelewski *et al.*, 2002; Sibille *et al.*, 2002). As with autosomal markers, microsatellites are favoured for forensic Y chromosome analysis and a number of suitable Y-STRs have been identified and validated for forensic use (Bosch *et al.*, 2002; Butler *et al.*, 2002; Redd *et al.*, 2002; Hall and Ballantyne, 2003; Johnson *et al.*, 2003; Hanson and Ballantyne, 2004; Schoske *et al.*, 2004) and a selection of them included into commercially available multiplexes (Shewale *et al.*, 2004; Krenke *et al.*, 2005; Mulero *et al.*, 2006).

Potentially the most valuable target markers for a diverse range of novel forensic molecular biology applications are single nucleotide polymorphisms (SNPs; see Chapter 6). These offer a range of forensic applications in traditional and novel areas and confer some particular advantages in comparison to STRs, including a low mutation rate (making SNPs highly suitable for kinship and/or pedigree analysis), amenability to high-throughput processing and automated data analysis, a shorter PCR amplicon size (assisting their ability to be multiplexed and making them good target loci for highly degraded samples), a vast abundance in the genome, and in some cases simplified interpretation (due to the absence of certain STR artifacts such as stutter). Single nucleotide polymorphisms are being investigated for use in forensics in both the identity testing and intelligence areas.

By virtue of the fact that there is greater allelic diversity at STR loci compared with SNPs, STRs have a profound advantage over SNPs in forensic identity testing. As a crude estimate, one would be required to type three to five SNP loci to discriminate between individuals at the same level as a single STR. This means that to approach the degree of certainty of the current STR kits up to 50 SNP loci would be needed, which presents a formidable technical challenge. In addition, changing routine target loci is undesirable, due largely to the significant investment in databases that has already occurred. In combination, these reasons make a universal change of DNA typing platform unlikely (Gill, 2001b; Gill *et al.*, 2004). Nonetheless, the recent development of more advanced SNP genotyping technologies, and the desirable properties of SNP loci, has seen a continued focus on developing highly informative SNP-based multiplexes for forensic identity testing (Inagaki *et al.*, 2004; Dixon *et al.*, 2006; Kidd *et al.*, 2006; Sanchez *et al.*, 2006).

Single nucleotide polymorphism markers in coding regions linked to physical or behavioural (personality-related) traits are also being researched for forensic purposes. This research aims to provide investigators with an inferred description of an offender, based on biological evidence recovered from a particular crime and subsequent DNA analysis. In one example researchers have described approaches for screening genetic mutations associated with the red-hair phenotype (Grimes *et al.*, 2001; Branicki *et al.*, 2006). A comprehensive candidate gene study for variable eye colour has also been conducted by an American company DNAPrint Genomics (Sarasota, FL, http://www.dnaprint.com) (Frudakis *et al.*, 2003a). On the basis of this research (Sturm and Frudakis, 2004) DNAPrint Genomics have developed and validated RETINOME™, a high-throughput genetic test for predicting human iris colour from DNA. A blind validation test of RETINOME™ on 65 individuals of greater than 80% European ancestry revealed that the test was 97% accurate in its predictions.

Other SNP-based techniques potentially enable the inference of biogeographical ancestry from a DNA sample. As SNPs can be found in areas of the genome subject to evolutionary-selective pressures, such as coding and regulatory regions of DNA, they can exhibit far greater allele and genotype frequency differences between different populations than other forensic loci. In 2003, Frudakis *et al.* developed a classifier for the SNP-based inference of ancestry (Frudakis *et al.*, 2003b). This research found that allele frequencies from 56 of the screened SNPs were notably different between groups of unrelated donors of Asian, African and European descent. Using this panel of 56 autosomal SNPs, Frudakis *et al.* report successful designation of the ancestral background of European, African and Asian donors with 99%, 98% and 100% accuracy, respectively. Applying a reduced panel of the 15 most informative SNPs the level of accuracy reduces to 98%, 91% and 97%, respectively (Frudakis *et al.*, 2003b). This work represents the most significant step towards the development of a DNA-based test for the inference of ancestry in a forensic setting and has led to the generation of a commercially available tool known as DNA Witness™ (DNA-Print Genomics, Sarasota, FL).

A significant amount of research effort has also been invested in the study of non-autosomal SNPs. This approach is commonplace in human migration studies, with a large body of work examining SNP haplotype diversity on the

Y chromosome (Underhill *et al.*, 2000, 2001; Jobling and Tyler-Smith, 2003) or mtDNA genome (Budowle *et al.*, 2004; Jobling *et al.*, 2004; Wilson and Allard, 2004). In the forensic context Y- or mtDNA-SNPs are also potential markers of biogeographical ancestry. They have often been preferred in this capacity as they can be locally customized and applied also to understand local population substructure, which in turn can support statistical interpretation models. Large-scale non-autosomal SNP multiplexes already exist (Sanchez *et al.*, 2003; Brion *et al.*, 2004, 2005; Coble *et al.*, 2004; Quintans *et al.*, 2004; Sanchez *et al.*, 2005) and population data and supporting information are readily available (YCC, 2002).

Commensurate with the advances in the molecular tools available to forensic scientists, the interpretation of DNA evidence has also had to develop considerably over recent years. Early in the history of forensic molecular biology this was an area of heated dispute (Lander, 1989; Lewontin and Hartl, 1991) requiring concerted efforts to address concerns of the scientific and legal community (National Academy of Sciences, 1996). Now there is a far greater depth of understanding and an important sub-discipline of the field has developed (Robertson and Vignaux, 1995; Evett and Weir, 1998; Aitken and Taroni, 2004; Balding, 2005; Buckleton *et al.*, 2005a). Nonetheless, each new molecular adaptation brings an associated requirement to reassess the weight or meaning of the outcomes statistically. Approaches are continually being refined to deal with routine complexities such as mixed profiles (Weir *et al.*, 1997; Curran *et al.*, 1999; Fukshansky and Bar, 2000; Bill *et al.*, 2005; Wolf *et al.*, 2005; Gill *et al.*, 2006a), partial profiles (Buckleton and Triggs, 2006) and relatedness (Ayres, 2000; Buckleton and Triggs, 2005). In addition, novel theory has been needed to assess results obtained from LCN approaches (Gill *et al.*, 2000), non-autosomal markers (Krawczak, 2001; Buckleton *et al.*, 2005b; Fukshansky and Bar, 2005; Wolf *et al.*, 2005), DNA database searches (Balding, 2002; Walsh and Buckleton, 2005), multi-trace cases (Aitken *et al.*, 2003), mass disasters (Brenner and Weir, 2003; Brenner, 2006) and so on.

From this summary we can distil the following trends that appear set to characterize future years. The addition of more routine markers, and the wider use of known ones, appears likely to continue. Testing platforms will increase in their overall efficiency and move closer to the goal of rapid, portable micro-devices. Taking the DNA science out of the laboratory is a move that could bring considerable advantage to many investigations but is also one with associated challenges. Progress will continue towards answering more diverse questions than 'who is the source of this DNA sample?'. There is almost limitless potential as to where this approach may lead as we unravel the full potential of information accessible via genetic testing. Of course we must observe that with this increased capability comes an associated increase in complexity. Scientists have the potential to step beyond the routinely applied testing regimes, but to do so they must understand the strengths and weaknesses of new approaches and, importantly, be equipped to deal with associated complexities such as the statistical assessment of outcomes. The forensic community must take ownership of this challenge and continue to ensure that proper validation, training and independent research occur. This will at times be awkward given the growing demands for all forms of DNA analysis and an increasingly commercialized operational environment. It will also be important to ensure appropriate management of expectations regarding emerging capabilities on the part of police, legal professionals and the general public.

1.3   Developments influencing law enforcement – operational impacts

The current environment where forensic molecular biology operates as a tool of the law enforcement community is starkly different to the mid-1980s, when its role in this context first began. This is unsurprising given the rapid evolution of the techniques, as described above. The most notable operational difference is the frequency of use of DNA evidence in criminal casework. Across the world the overall number of cases submitted annually for DNA analysis has increased by many fold. In the UK the average annual inclusion of crime samples onto the national DNA database (NDNAD™) increased from 14 644 for the period 1995–2000 to 59 323 for the period 2000–2005. In Canada, 7052 crime samples were added to the national DNA databank in 2005 compared with 816 in 2000. In NSW (the most populous State of Australia) the annual DNA case submissions have risen from 1107 in 1998 to 10 146 in 2005.

The major driver of this change in case volume has been the global implementation of forensic DNA databases. Forensic DNA databases have altered the landscape of the criminal justice system and irrevocably re-shaped the field of forensic science. Their growth has been rapid with millions of STR profiles now held from convicted offenders, suspects and unsolved crimes (Table 1.1). Links provided through DNA database searches have contributed valuable

Table 1.1   Size and effectiveness of major national DNA databases

	Database and date				
	UK Feb. 2006	Europe Dec. 2005	USA Apr. 2006	Canada May 2006	New Zealand Apr. 2005
Total profiles	3 693 494	987 671	3 275 710	123 603	63 678
Offender profiles	3 406 488	772 355	3 139 038	94 999	54 159
Crime scene profiles	287 006	215 316	136 672	28 604	9 419
Investigations aided	721 495	116 057	34 193	5 963	2 451

Source: Publicly available figures on the Internet.

intelligence to hundreds of thousands of police investigations. Often links are provided for crimes that are notoriously difficult to resolve, such as burglary and vehicle theft.

Along with the increase in case volume that has been catalysed in part by the introduction of DNA databases, there has also been an alteration to the types of crimes and evidence submitted for biological analysis (see Chapter 11). In the 1980s and 1990s DNA profiling was primarily applied to serious crimes. Nowadays, however, forensic molecular biology contributes to the investigation of a broader spectrum of crimes. Data from the NSW State Forensic DNA Laboratory over the period 1998–2005 show a clear pattern of decrease in the proportion of cases from serious crime categories and an increase in the proportion of cases submitted from volume crime categories (Figure 1.2). The change in the case submission profile, that is, the proportions of different case types submitted for analysis, occurred from 2001 forward. This was the beginning of DNA database operations.

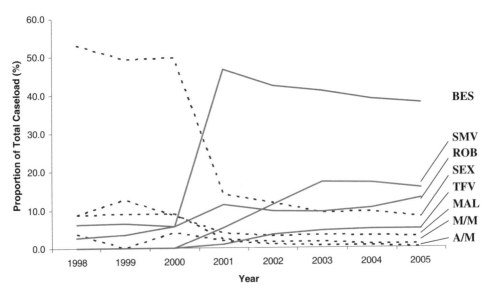

Figure 1.2 Changing case submission trends at a State Forensic DNA Laboratory in Australia. Case categories where there was a decrease in proportional submissions are represented by the dashed lines. The actual categories are Sexual Assault (SEX; 53.0% to 8.0%), Murder & Manslaughter (M/M; 8.7% to 1.0%), Malicious Wounding (MAL; 8.9% to 2.9%) and Attempt Murder (A/M; 3.6% to 0.3%). The case categories where there was an increase in proportional submissions are represented by the solid lines. These include Break, Enter and Steal (BES; 2.7% to 37.8%), Stolen Vehicle (SMV; 0.0% to 15.6%), Robbery (ROB; 6.3% to 12.8%) and Theft from Vehicle (TFV; 0.0% to 4.9%)

Table 1.2 Changing evidence types analysed by the NSW State Forensic DNA Laboratory between 2001 and 2005

Sample type	Proportion of total (%)				
	2001	2002	2003	2004	2005
Blood	44.7	49.1	38.4	33.8	26.5
Bone	0.0	0.1	0.2	0.1	0.1
Cigarette butt	4.5	8.0	10.9	8.4	9.1
Hair	10.0	3.1	2.2	2.0	1.3
Miscellaneous[a]	30.9	33.4	41.1	50.4	57.3
Semen	8.3	4.6	5.7	3.9	4.4
Fingernail scrapings	1.7	1.8	1.5	1.3	1.2

[a] Includes mostly trace DNA samples.

The changing nature of the case submission profile in forensic molecular biology laboratories has been accompanied by a changing evidence type away from traditional templates such as blood and semen to more discrete evidence types such as trace DNA and discarded items (including drinking containers and cigarette butts). This is again illustrated in data from the NSW State Forensic DNA Laboratory from between 2001 and 2005 (Table 1.2).

As well as having a profound effect on the number and types of cases submitted for DNA analysis, forensic DNA databases have also catalysed a re-think of the role that forensic evidence can play in the investigative process. Traditionally forensic DNA evidence has been thought of as information for the use of the court. This focus sees the scientist retrospectively attempting to obtain results for a given case to assist in the resolution of a single crime. The primary focus, therefore, is towards crime-solving rather than crime reduction or prevention. For some years now, policing strategy has evolved from a traditional focus on capturing or incarcerating offenders towards a more holistic understanding of crimes and criminals and a prevention-based approach to law enforcement. Forensic science has not contributed greatly to changes in policing and crime management strategies, although lately there has been a move to embrace scientific advances under the concept of intelligence-led policing (Smith, 1997; Thompson and Gunn, 1998; Gunn, 2003; Tilley, 2003). Forensic molecular biology clearly has a role to play in the generation of law enforcement intelligence products (Walsh et al., 2002), particularly when one considers the potential of combining rapid, portable DNA typing with the use of DNA databases or phenotypic inferences about an offender. As such it has the potential to play a more proactive role in broader-scale crime investigation.

Achieving this requires shifting the philosophical mind-set of forensic practitioners, understanding where, when and how forensic science data can be useful in an intelligence context, and designing systems capable of relaying findings in 'real-time'. A number of approaches have begun to emerge that embrace this operational strategy. Some remain ill-informed and are based around centralized database creation. Other more successful examples create meaningful forensic intelligence and combine it with investigative and crime analysis tools (Ribaux and Margot, 1999, 2003).

In summary there has been a trend towards greater use of DNA, across a more diverse range of cases and in a more intelligence-based context or framework. It is important to note, however, that these developments (and the DNA databases that have predominantly catalysed them) remain at a preliminary stage. Standards and approaches still vary enormously between jurisdictions and in future there may be continued moves towards greater harmonization. Undoubtedly there will be progress towards greater cross-jurisdictional exchange of DNA information, possibly facilitated at the level of organizations such as Interpol. Managing this era of wider national and international use of forensic DNA profile information will be challenging as, through these developments, the science we apply moves increasingly into the public and political realm.

1.4   Developments influencing the justice system – socio-legal impacts

Practical and philosophical aspects of the legal system have been impacted by developments in the field of forensic molecular biology. Practical issues emanate from the construction of laws that regulate the collection of DNA material from persons associated with the justice system, and the subsequent use of any DNA-related evidence in our courts. These are flanked by important philosophical considerations in areas such as social justice, ethics and privacy.

From the time of its first introduction the courts have had mixed experiences with the presentation of forensic DNA evidence. Scientists, lawyers, judges and jurors have battled to come to terms with this new forensic application of a complex scientific technique. Initially complicating matters further was the public fanfare that accompanied early DNA successes, creating an aura of scientific certainty around the technology. Whilst forensic molecular biology is a powerful means of identification, this sort of misrepresentation in the public arena can create unrealistic perceptions of its capability. At a relatively early stage, the admissibility of DNA evidence in criminal trials was successfully challenged in the United States (*People v Castro*, 1989, 545 NYS 2d 985) and elsewhere (*R v Tran*, 1990, 50 A Crim R 233; *R v Lucas*, 1992, 2 VR 109; *R v Pengelly*, 1992, 1 NZLR 545). Many of the issues upon which early challenges were mounted were the subject of conflict in the scientific community at the time (Lander, 1989; Chakraborty and Kidd, 1991; Lewontin and Hartl, 1991). The scrutiny of the legal system in these instances must be seen to have been strongly positive as it brought about further refinement and validation of the forensic DNA methodology and the implementation of structures to regulate quality assurance. In recent times challenges to DNA admissibility are rarely successful, as, in general terms, the science has reached the important point of being accepted practice. This is not to say that legal scrutiny has abated entirely, rather, if anything, legal challenges have evolved in their complexity along with the evolution of the technology itself. Instead of focusing on general issues, it is now specialized components of the analytical or interpretative process that have become the subject of questioning.

Widening the use of forensic DNA evidence and implementing forensic DNA databases have required the formulation of specific statutes. The enactment of the DNA-based laws has generated considerable commentary in the legal literature regarding the process of enactment and the final constitution of the laws themselves. In most cases the legal discussion is critical, suggesting that the passage of legislation was hurried and justified under the populist appeal of 'law and order' politics. Others fear that compulsory acquisition of genetic material by the state represents an encroachment into previously sacred territory of criminal law and a diminution of basic rights such as the right to silence and the right against self-incrimination. Many commentators express concern over the mass storage of human genetic information and the associated potential risk of future misuse.

Overall, the practical developments at the intersection of forensic molecular biology and the legal system have progressed from general issues (such as whether DNA testing has a place in the CJS at all) to specific refinements (such as how DNA evidence was obtained and tested). Again it is important to note that this period of interplay between the science of DNA testing and the regulation of legal sector is a relatively recent phenomenon and is bound to evolve considerably, even in the short term. The balance that is sought relates to attempting to achieve maximal effectiveness from the use of DNA, whilst minimizing the incursion into a person's basic civil and legal rights. In different cases, different countries and at different times striking this balance can be influenced by external pressures. Recently, for example, heightened anxiety around terrorism has seen governments override individual rights in favor of more expansive investigative powers. The use of forensic molecular biology is linked to many of these broad socio-legal issues.

1.5   Summary

The field of forensic molecular biology has entered a period of development where more genetically diverse applications are emerging and are able to be delivered by a more responsive and technically advanced platform. There is also an emphasis for forensic DNA outcomes to be delivered as intelligence products as well as evidence for the court. This allows it to take on a more purposeful role in the investigative phase of the process alongside other items of forensic or non-forensic intelligence. Understandably, these enhancements in capability are continuing to drive a great demand to access and utilize this technology. So far this demand has, in many cases, outstripped the ability of forensic laboratories to cope, and case backlogs are commonplace in many jurisdictions.

These trends signal the beginning of an exciting era for forensic molecular biology. As forensic professionals, however, we must remember that adding these dimensions to our capability also adds to our overall onus of responsibility. Such changes require a continual broadening of our outlook and expertise. Also, due to the emotional and social stakes that exist in the criminal justice system, many developments in our field are somewhat double-edged: able to be viewed either positively as strengthening our ability to fight against crime or terrorism or negatively as examples of an increasing loss of civil liberties and greater surveillance by the state on her citizens. Whilst forensic scientists remain impartial players in this environment, it is important that we do not extricate ourselves from this debate that essentially defines how and to what end our scientific endeavour is applied. Of paramount importance is that, across all techniques within our field, we continue to ensure the quality of our scientific outcomes. By doing this our work will remain an objective and reliable component of the criminal justice system.

1.6   References

Abaz, J., Walsh, S.J., Curran, J.M., Moss, D.S., Cullen, J.R., Bright, J., Crowe, G.A., Cockerton, S.L. and Power, T.E.B. (2002) Comparison of variables affecting the recovery of DNA from common drinking containers. *Forensic Sci. Int.* **126**: 233–240.

Aitken, C.G.G. and Taroni, F. (2004) *Statistics and the Evaluation of Evidence for Forensic Scientists* (2nd edn), John Wiley & Sons, Chichester.

Aitken, C.G.G., Taroni, F. and Garbolino, P. (2003) A graphical model for the evaluation of cross-transfer evidence in DNA profiles. *Theor. Pop. Biol.* **63**: 179–190.

Andelinovic, S., Sutlovic, D., Erceg Ivkosic, I., Skaro, V., Ivkosic, A., Paic, F., Rezic, B., Definis-Gojanovic, M. and Primorac, D. (2005) Twelve-year experience in identification of skeletal remains from mass graves. *Croat. Med. J.* **46**: 530–539.

Anderson, T.D., Ross, J.P., Roby, R.K., Lee, D.A. and Holland, M.M. (1999) A validation study for the extraction and analysis of DNA from human nail material and its application to forensic casework. *J. Forensic Sci.* **44**: 1053–1056.

Ayres, K.L. (2000) Relatedness testing in subdivided populations. *Forensic Sci. Int.* **114**: 107–115.

Balding, D.J. (2002) The DNA database search controversy. *Biometrics* **58**: 241–244.

Balding, D.J. (2005) *Weight-of-Evidence for Forensic DNA Profiles*, John Wiley & Sons, Hoboken, New Jersey.

Bender, K., Schneider, P.M. and Rittner, C. (2000) Application of mtDNA sequence analysis in forensic casework for the identification of human remains. *Forensic Sci. Int.* **113**: 103–107.

Bill, M., Gill, P., Curran, J.M., Clayton, T., Pinchin, R., Healy, M. and Buckleton, J.S. (2005) PENDULUM – a guideline-based approach to the interpretation of STR mixtures. *Forensic Sci. Int.* **148**: 181–189.

Bosch, E., Lee, A.C., Calafell, F., Arroyo, E., Henneman, P., de Knijff, P. and Jobling, M.A. (2002) High resolution Y chromsome typing: 19 STRs amplified in three multiplex reactions. *Forensic Sci. Int.* 125: 42–51.

Branicki, W., Kupiec, T., Wolanska-Nowak, P. and Brudnik, U. (2006) Determination of forensically relevant SNPs in the MC1R gene. *Int. Congr. Ser.* 1288: 816–818.

Brenner, C.H. (2006) Some mathematical problems in the DNA identification of victims in the 2004 tsunami and similar mass fatalities. *Forensic Sci. Int.* 157: 172–180.

Brenner, C.H. and Weir, B.S. (2003) Issues and strategies in the DNA identification of World Trade Center victims. *Theor. Pop. Biol.* 63: 173–178.

Bright, J.A. and Petricevic, S.F. (2004) Recovery of trace DNA and its application to DNA profiling of shoe insoles. *Forensic Sci. Int.* 145: 7–12.

Brion, M., Sanchez, J.J., Balogh, K., Thacker, C., Blanco-Verea, A., Borsting, C., Stradmann-Bellinghausen, B., Bogus, M., Syndercombe-Court, D., Schneider, P.M., Carracedo, A. and Morling, N. (2005) Introduction of an single nucleodite polymorphism-based 'Major Y-chromosome haplogroup typing kit' suitable for predicting the geographical origin of male lineages. *Electrophoresis* 26: 4411–4420.

Brion, M., Sobrino, B., Blanco-Verea, A., Lareu, M.V. and Carracedo, A. (2004) Hierachical analysis of 30 Y-chromosome SNPs in European populations. *Int. J. Legal Med.* 119: 10–15.

Buckleton, J.S. and Triggs, C.M. (2005) Relatedness and DNA: are we taking it seriously enough? *Forensic Sci. Int.* 152: 115–119.

Buckleton, J.S. and Triggs, C.M. (2006) Is the 2p rule always conservative? *Forensic Sci. Int.* 159: 206–209.

Buckleton, J.S., Triggs, C.M. and Walsh, S.J. (2005a) *Forensic DNA Evidence Interpretation*, CRC Press, Boca Raton, FL.

Buckleton, J.S., Walsh, S.J. and Harbison, S.A. (2005b) Nonautosomal forensic markers. In: *Forensic DNA Evidence Interpretation* (J.S. Buckleton, C.M. Triggs and S.J. Walsh, eds), CRC Press, Boca Raton, FL, pp. 299–339.

Budjimila, Z.M., Prinz, M.K., Zelson-Mundorff, A., Wiersema, J., Bartelink, E., MacKinnon, G., Nazzaruolo, B.L., Estacio, S.M., Hennessey, M.J. and Shaler, R.C. (2003) World Trade Center human identification project: experiences with individual body identification cases. *Croat. Med. J.* 44: 259–263.

Budowle, B., Bieber, F.R. and Eisenberg, A.J. (2005) Forensic aspects of mass disasters: strategic considerations for DNA-based human identification. *Legal Med.* 7: 230–245.

Budowle, B., Planz, J.V., Campbell, R.S. and Eisenberg, A.J. (2004) Single nucleotide polymorphisms and microarray technology in forensic genetics – development and application to mitochondrial DNA. *Forensic Sci. Rev.* 16: 21–36.

Butler, J.M., Li, J., Shaler, T.A., Monforte, J.A. and Becker, C.H. (1998) Reliable genotyping of short tandem repeat loci without an alleleic ladder using time-of-flight mass spectrometry. *Int. J. Legal Med.* 112: 45–49.

Butler, J.M., Schoske, R., Vallone, P.M., Kline, M.C., Redd, A.J. and Hammer, M.F. (2002) A novel multiplex for simultaneous amplification of 20 Y chromosome STR markers. *Forensic Sci. Int.* 129: 10–24.

Butler, J.M., Shen, Y. and McCord, B.R. (2003) The development of reduced size STR amplicons as tools for analysis of degraded DNA. *J. Forensic Sci.* 48: 1054–1064.

Cann, R.L., Stoneking, M. and Wilson, A.C. (1987) Mitochondrial DNA and human evolution. *Nature* 325: 31–36.

Chakraborty, R. and Kidd, K.K. (1991) The utility of DNA typing in forensic work. *Science* **254**: 1735–1744.

Coble, M.D. and Butler, J.M. (2005) Characterization of new miniSTR loci to aid analysis of degraded DNA. *J. Forensic Sci.* **50**: 43–53.

Coble, M.D., Just, R.S., O'Callaghan, J.E., Letmanyi, I.H., Peterson, C.T., Irwin, J.A. and Parsons, T.J. (2004) Single nucleotide polymorphisms over the entire mtDNA genome that increase the power of forensic testing in Caucasians. *Int. J. Legal Med.* **118**: 137–146.

Collins, P., Roby, R.K., Lori, H., Leibelt, C., Shadravan, F., Bozzini, M.L. and Reeder, D. (2000) Validation of the AmpFlSTR Identifiler PCR amplification kit. In *Proceedings of 11th International Symposium on Human Identification*, Biloxi, Missouri, September 2000. Available at http://www.promega.com/geneticidproc/ussymp11proc/content/collins.pdf

Curran, J.M., Triggs, C.M., Buckleton, J.S. and Weir, B.S. (1999) Interpreting DNA mixtures in structured populations. *J. Forensic Sci.* **44**: 987–995.

Dettlaff-Kakol, A. and Pawlowski, R. (2002) The first Polish DNA 'manhumnt' – an application of Y-chromosome STR's. *Int. J. Legal Med.* **116**: 289–291.

Dixon, L.A., Koumi, P. and Gill, P. (2006) Development of an autosomal SNP multiplex containing 20 SNP loci plus amelogenin. *Int. Congr. Ser.* **1288**: 31–33.

Dziegelewski, M., Simich, J.P. and Rittenhouse-Olsen, K. (2002) Use of a Y chromosome probe as an aid in the forensic proof of sexual assault. *J. Forensic Sci.* **47**: 601–604.

Edson, S.M., Ross, J.P., Coble, M.D., Parsons, T.J. and Barritt, S.M. (2004) Naming the dead – confronting the realities of rapid identification of degraded skeletal remains. *Forensic Sci. Rev.* **16**: 63–90.

Evett, I.W. and Weir, B.S. (1998) *Interpreting DNA Evidence*, Sinauer Associates, Sunderland, MA.

Fregeau, C.J., Leclair, B., Bowen, K., Porelle, F. and Fourney, R.M. (2003) The National DNA Data Bank of Canada – a laboratory bench retrospective on the first year of operation. *Int. Congr. Ser.* **1239**: 621–625.

Frudakis, T., Thomas, M., Gaskin, Z., Venkateswarlu, K., Chandra, K., Ginjupalli, S., Gunturi, S., Natrajan, S., Ponnuswamy, V. and Ponnuswamy, K. (2003a) Sequences associated with human iris pigmentation. *Genetics* **165**: 2071–2083.

Frudakis, T., Venkateswarlu, K., Thomas, M.J., Gaskin, Z., Ginjupalli, S., Gunturi, S., Ponnuswamy, V., Natarajan, S. and Nachimuthu, P.K. (2003b) A classifier for the SNP-based inference of ancestry. *J. Forensic Sci.* **48**: 771–782.

Fukshansky, N. and Bar, W. (2000) Biostatisitcs for mixed stains: the case of tested relatives of a non-tested suspect. *Int. J. Legal Med.* **114**: 78–82.

Fukshansky, N. and Bar, W. (2005) DNA mixtures: biostatistics for mixed stains with haplotypic genetic markers. *Int. J. Legal Med.* **119**: 285–290.

Gill, P. (2001a) Application of low copy number DNA profiling. *Croat. Med. J.* **42**: 229–232.

Gill, P. (2001b) An assessment of the utility of single nucleotide polymorphisms (SNPs) for forensic purposes. *Int. J. Legal Med.* **114**: 204–210.

Gill, P. (2002) Role of short tandem repeat DNA in forensic casework in the UK – past, present and future perspectives. *BioTechniques* **32**: 366–385.

Gill, P., Brenner, C., Brinkmann, B., Budowle, B., Carracedo, A., Jobling, M.A., de Knijff, P., Kayser, M., Krawczak, M., Mayr, W.R., Morling, N., Olaisen, B.,

Pascali, V., Prinz, M., Roewer, L., Schneider, P.M., Sajantila, A. and Tyler-Smith, C. (2001) DNA Commission of the International Society of Forensic Genetics: recommendations on forensic analysis using Y-chromosome STRs. *Forensic Sci. Int.* **124**: 5–10.

Gill, P., Brenner, C.H., Buckleton, J.S., Carracedo, A., Krawczak, M., Mayr, W.R., Morling, N., Prinz, M., Schneider, P.M. and Weir, B.S. (2006a) DNA Commission of the International Society of Forensic Genetics: Recommendations on the interpretation of mixtures. *Forensic Sci. Int.* **160**: 90–101.

Gill, P., Curran, J.M. and Elliot, K. (2005) A graphical simulation model of the entire DNA process associated with the analysis of short tandem repeat loci. *Nucleic Acids Res.* **33**: 632–643.

Gill, P., Ivanov, P.L., Kimpton, C., Piercy, R., Benson, N., Tully, G., Evett, I.W., Hagelberg, E. and Sullivan, K. (1994) Identification of the remains of the Romanov family by DNA analysis. *Nature Genet.* **6**: 130–135.

Gill, P., Kirkham, A. and Curran, J.M. (2006b) *LoComatioN*: A software tool for the analysis of low copy number profiles. *Forensic Sci. Int.* [Epub ahead of print] doi: 10.1016/j.forsciint.2006.04.016.

Gill, P., Werrett, D.J., Budowle, B. and Guerrieri, R.A. (2004) An assessment of whether SNPs will replace STRs in national DNA databases – Joint considerations of the DNA working group of the European Network of Forensic Science Institutes (ENFSI) and the Scientific Working Group on DNA Analysis Methods (SWGDAM). *Sci. Justice* **44**: 51–53.

Gill, P., Whitaker, J.P., Flaxman, C., Brown, N. and Buckleton, J.S. (2000) An investigation of the rigor of interpretation rules for STR's derived from less than 100 pg of DNA. *Forensic Sci. Int.* **112**: 17–40.

Grimes, E.A., Noakes, P.J., Dixon, L. and Urquhart, A. (2001) Sequence polymorphism in the human melanocortin 1 receptor gene as an indicator of the red hair phenotype. *Forensic Sci. Int.* **122**: 124–129.

Gunn, B. (2003) An intelligence-led approach to policing in England and Wales and the impact of developments in forensic science. *Aust. J. Forensic Sci.* **35**: 149–160.

Gusmao, L., Butler, J.M., Carracedo, A., Gill, P., Kayser, M., Mayr, W.R., Morling, N., Prinz, M., Roewer, L., Tyler-Smith, C. and Schneider, P.M. (2006) DNA Commission of the International Society of Forensic Genetics (ISFG): An update of the recommendations on the use of Y-STRs in forensic analysis. *Forensic Sci. Int.* **157**: 187–197.

Gusmao, L. and Carracedo, A. (2003) Y chromosome-specific STRs. *Profiles DNA* **5**: 3–6.

Hall, A. and Ballantyne, J. (2003) The development of an 18-locus Y-STR system for forensic casework. *Anal. Bioanal. Chem.* **376**: 1234–1246.

Hanson, E.K. and Ballantyne, J. (2004) A highly discriminating 21 locus 'megaplex' system designed to augment the minimal haplotype loci for forensic casework. *J. Forensic Sci.* **49**: 40–51.

Harbison, S.A., Petricevic, S.F. and Vintiner, S.K. (2003) The persistence of DNA under fingernails following submersion in water. *Int. Congr. Ser.* **1239**: 809–813.

Holland, M.M., Cave, C.A., Holland, C.A. and Bille, T.W. (2003) Development of a quality, high throughput DNA analysis procedure for skeletal samples to assist with the identification of victims from the world trade center attacks. *Croat. Med. J.* **44**: 264–272.

Hopwood, A., Fox, R., Round, C., Tsang, C., Watson, S., Rowlands, E., Titmus, A., Lee-Edghill, J., Cursiter, L., Proudlock, J., McTernan, C., Grigg, K., Thornton, L. and Kimpton, C. (2006) Forensic response vehicle: Rapid analysis of evidence at the crime scene. *Int. Congr. Ser.* **1288**: 639–641.

Huffine, E., Crews, J., Kennedy, B., Bomberger, K. and Zinbo, A. (2001) Mass identification of persons missing from the break-up of the former Yugoslavia: structure, function, and role of International Commission on Missing Persons. *Croat. Med. J.* **42**: 271–275.

Inagaki, S., Yamamoto, Y., Doi, Y., Takata, T., Ishikawa, T., Imabayashi, K., Yoshitome, K., Miyaishi, S. and Ishizu, H. (2004) A new 39-plex analysis method for SNPs including 15 blood group loci. *Forensic Sci. Int.* **144**: 45–57.

Ivanov, P.L., Wadhams, M.J., Roby, R.K., Holland, M.M., Weedn, V.W. and Parsons, T.J. (1996) Mitochondrial DNA sequence heteroplasmy in the Grand Duke of Russia Georgij Romanov establishes the authenticity of the remains of Tsar Nicholas II. *Nature Genet.* **12**: 417–420.

Jobling, M.A., Hurles, M.E. and Tyler-Smith, C. (2004) *Human Evolutionary Genetics*, Garland Science, New York.

Jobling, M.A., Pandya, A. and Tyler-Smith, C. (1997) The Y chromosome in forensic analysis and paternity testing. *Int. J. Legal Med.* **110**: 118–124.

Jobling, M.A. and Tyler-Smith, C. (2003) The human Y chromosome: an evolutionary marker comes of age. *Nature Rev. Genet.* **4**: 598–612.

Johnson, C.L., Warren, J.H., Giles, R.C. and Staub, R.W. (2003) Validation and uses of a Y-chromosome STR 10-plex for forensic and paternity laboratories. *J. Forensic Sci.* **48**: 1260–1268.

Kayser, M., Caglia, A., Corach, D., Fretwell, N., Gehrig, C., Graziosi, G., Heidorn, F., Herrmann, S., Herzog, B., Hidding, M., Honda, K., Jobling, M.A., Krawczak, M., Leim, K., Meuser, S., Meyer, E., Oesterreich, W., Pandya, A., Parson, W., Penacino, G., Perez-Lezaun, A., Piccinini, A., Prinz, M., Schmitt, C., Schneider, P.M., Szibor, R., Teifel-Greding, J., Weichhold, G., de Knijff, P. and Roewer, L. (1997) Evaluation of Y-chromosomal STRs: a multicenter study. *Int. J. Legal Med.* **110**: 125–133.

Kidd, K.K., Pakstis, A.J., Speed, W.C., Grigorenko, E.L., Kajuna, S.L.B., Karoma, N.J., Kungulilo, S., Kim, J.-J., Lu, R.-B., Odunsi, A., Okonofua, F., Parnas, J., Schulz, L.O., Zhukova, O.V. and Kidd, J.R. (2006) Developing a SNP panel for forensic identification of individuals. *Forensic Sci. Int.* [Epub ahead of print] doi: 10.1016/j.forscient.2006.04.016.

Krawczak, M. (2001) Forensic evaluation of Y-STR haplotype matches: a comment. *Forensic Sci. Int.* **118**: 114–115.

Krenke, B., Tereba, A., Anderson, S.J., Buel, E., Culhane, S., Finis, C.J., Tomsey, C.S., Zachetti, J.M., Masibay, A., Rabbach, D.R., Amiott, E.A. and Sprecher, C.J. (2002) Validation of a 16-locus fluorescent multiplex system. *J. Forensic Sci.* **47**: 773–785.

Krenke, B.E., Viculis, L., Richard, M.L., Prinz, M., Milne, S.C., Ladd, C., Gross, A.M., Gornall, T., Frappier, J.R., Eisenberg, A.J., Barna, C., Aranda, X.G., Adamowicz, M.S. and Budowle, B. (2005) Validation of male-specific, 12-locus fluorescent short tandem repeat (STR) multiplex. *Forensic Sci. Int.* **151**: 111–124.

Lander, E.S. (1989) DNA fingerprinting on trial. *Nature* **339**: 501.

Lewontin, R.C. and Hartl, D.L. (1991) Population genetics in forensic DNA typing. *Science* **254**: 1745–1750.

Lorente, J.A., Entrala, C., Alvarez, J.C., Lorente, M., Arce, B., Heinrich, B., Carrasco, F., Budowle, B. and Villanueva, E. (2002) Social benefits on non-criminal genetic databases: missing persons and human remains identification. *Int. J. Legal Med.* **116**: 187–190.

Lorente, M., Entrala, C., Lorente, J., Alvarez, J.C., Villanueva, E. and Budowle, B. (1998) Dandruff as a potential source of DNA in forensic casework. *J. Forensic Sci.* **43**: 648–656.

Mitnik, L., Carey, L., Burger, R., Desmarais, S., Koutny, L., Wernet, O., Matsudaira, P. and Ehrlich, D. (2002) High-speed analysis of multiplexed short tandem repeats with and electrophoretic microdevice. *Electrophoresis* **23**: 719–726.

Mulero, J.J., Chang, C.W., Calandro, L.M., Green, R.L., Li, Y., Johnson, C.L. and Hennessy, L.K. (2006) Development and validation of the AmpFlSTR Yfiler PCR amplification kit: a male specific, single amplification 17 Y-STR multiplex system. *J. Forensic Sci.* **51**: 64–75.

National Academy of Sciences (1996) *National Research Council Report: The Evaluation of Forensic DNA Evidence*, United States National Academy of Sciences, Washington, DC.

Perlin, M.W. and Szabady, B. (2001) Linear mixture analysis: a mathematical approach to resolving mixed DNA samples. *J. Forensic Sci.* **46**: 1372–1378.

Pfeiffer, H., Benthaus, S., Rolf, B. and Brinkmann, B. (2003) The Kaiser's tooth. *Int. J. Legal Med.* **117**: 118–120.

Quintans, B., Alvarez-Iglesias, V., Salas, A., Phillips, C., Lareu, M.V. and Carracedo, A. (2004) Typing of mitochondrial DNA coding region SNPs of forensic and anthropological interest using SNaPshot minisequencing. *Forensic Sci. Int.* **140**: 251–257.

Redd, A.J., Agellon, A.B., Kearney, V.A., Contreras, V.A., Karafet, T., Park, H., de Knijff, P., Butler, J.M. and Hammer, M.F. (2002) Forensic value of 14 novel STRs on the human Y chromosome. *Forensic Sci. Int.* **130**: 97–111.

Ribaux, O. and Margot, P. (1999) Inference structures for crime analysis and intelligence: the example of burglary using forensic science data. *Forensic Sci. Int.* **100**: 193–210.

Ribaux, O. and Margot, P. (2003) Case based reasoning in criminal intelligence using forensic case data. *Sci. Justice* **43**: 135–143.

Robertson, B. and Vignaux, G.A. (1995) *Interpreting Evidence: Evaluating Forensic Science in the Courtroom*, John Wiley & Sons, Chichester.

Roby, R.K. (2002) Automation for forensic mitochondrial DNA analysis. In *Proceedings of 5th Annual DNA Forensics Conference*, Washington, DC.

Sanchez, J.J., Borsting, C., Hallenberg, C., Buchard, A., Hernandez, A. and Morling, N. (2003) Multiplex PCR and minisequencing of SNPs – a model with 35 Y chromosome SNPs. *Forensic Sci. Int.* **137**: 74–84.

Sanchez, J.J., Borsting, C. and Morling, N. (2005) Typing of Y chromosome SNPs with multiplex PCR methods. *Methods Mol. Biol.* **297**: 209–228.

Sanchez, J.J., Phillips, C., Borsting, C., Balogh, K., Bogus, M., Fondevila, M., Harrison, C.D., Musgrave-Brown, E., Salas, A., Syndercombe-Court, D., Schneider, P.M., Carracedo, A. and Morling, N. (2006) A multiplex assay with 52 single nucleotide polymorphisms for human identification. *Electrophoresis* **27**: 1713–1724.

Schoske, R., Vallone, P.M., Kline, M.C., Redman, J.W. and Butler, J.M. (2004) High-throughput Y-STR typing of U.S. populations with 27 regions of the Y chromosome using two multiplex PCR assays. *Forensic Sci. Int.* **139**: 107–121.

Shewale, J.G., Nasir, H., Schneida, E., Gross, A.M., Budowle, B. and Sinha, S.K. (2004) Y-chromosome STR system, Y-PLEX™ 12, for forensic casework: development and validation. *J. Forensic Sci.* **49**: 1278–1290.

Sibille, I., Duverneuil, C., Lorin de la Grandmaison, G., Guerrouache, K., Teisiere, F., Durigon, M. and de Mazancourt, P. (2002) Y-STR DNA amplification as biological evidence in sexually assaulted female victims with no cytological detection of spermatozoa. *Forensic Sci. Int.* **125**: 212–216.

Smith, A. (1997) *Intelligence Led Policing; International Perspectives on Policing in the 21st Century*, International Association of Law Enforcement Intelligence Analysts, Lawrenceville, CA.

Sturm, R. and Frudakis, T. (2004) Eye colour: portals into pigmentation genes and ancestry. *Trends Genet.* **20**: 327–332.

Sweet, D. and Hildebrand, D. (1999) Saliva from cheese bite yields DNA profile of a burglar: a case report. *Int. J. Legal Med.* **112**: 201–203.

Thompson, J. and Gunn, B. (1998) Tomorrow's world. *Policing Today* **4**: 12–13.

Tilley, N. (2003) *Problem-Oriented Policing, Intelligence-Led Policing and the National Intelligence Model*, Report of the Jill Dando Institute of Crime Science, Crime Science: Short Report Series, University College London, UK.

Tully, G., Bär, W., Brinkmann, B., Carracedo, A., Gill, P., Morling, N., Parson, W. and Schneider, P.M. (2001) Considerations of the European DNA Profiling (EDNAP) group on the working practices, nomenclature and interpretation of mitochondrial DNA profiles. *Forensic Sci. Int.* **124**: 83–91.

Underhill, P.A., Passarino, G., Lin, A.A., Shen, P., Mirazon Lahr, M., Foley, R.A., Oefner, P.J. and Cavalli-Sforza, L.L. (2001) The phylogeography of Y chromosome binary haplotypes and the origins of modern human populations. *Ann. Hum. Genet.* **65**: 43–62.

Underhill, P.A., Shen, P., Lin, A.A., Jin, L., Passarino, G., Yang, W.-H., Kauffman, E., Bonne-Tamir, B., Bertranpetit, J., Francalacci, P., Ibrahim, M., Jenkins, T., Kidd, J.R., Medhi, S.Q., Seielstad, M.T., Wells, R.S., Piazza, A., Feldman, M.W., Cavalli-Sforza, L.L. and Oefner, P.J. (2000) Y chromosome sequence variation and the history of human populations. *Nature Genet.* **26**: 358–361.

Van Oorschot, R.A. and Jones, M. (1997) DNA fingerprints from fingerprints. *Nature* **387**: 767.

Varlaro, J. and Duceman, B. (2002) Dealing with increasing casework demands for DNA analysis. *Profiles DNA* **5**: 3–6.

Vastag, B. (2002) Out of tragedy, identification innovation. *J. Am. Med. Assoc.* **288**: 1221–1223.

Walsh, S.J. and Buckleton, J.S. (2005) DNA intelligence databases. In: *Forensic DNA Evidence Interpretation*, (J.S. Buckleton, C.M. Triggs, and S.J. Walsh, eds). CRC Press, Boca Raton, FL, pp. 439–469.

Walsh, S.J., Moss, D.S., Kleim, C. and Vintiner, G.M. (2002) The collation of forensic DNA case data into a multi-dimensional intelligence database. *Sci. Justice* **42**: 205–214.

Webb, L.G., Egan, S.E. and Turbett, G.R. (2001) Recovery of DNA for forensic analysis from lip cosmetics. *J. Forensic Sci.* **46**: 1474–1479.

Weir, B.S., Triggs, C.M., Starling, L., Stowell, L.I., Walsh, K.A.J. and Buckleton, J.S. (1997) Interpreting DNA mixtures. *J. Forensic Sci.* **42**: 213–222.

Whitaker, J.P., Cotton, E.A. and Gill, P. (2001) A comparison of the characteristics of profiles produced with the AMPFlSTR®SGM Plus™ multiplex system for both standard and low copy number (LCN) STR DNA analysis. *Forensic Sci. Int.* **123**: 215–223.

Wiegand, P. and Kleiber, M. (1997) DNA typing of epithelial cells after strangulation. *Int. J. Legal Med.* **110**: 181–183.

Wiegand, P. and Kleiber, M. (2001) Less is more – length reduction of STR amplicons using redesigned primers. *Int. J. Legal Med.* **114**: 285–287.

Wilson, M.R. and Allard, M.W. (2004) Phylogentics and mitochondrial DNA. *Forensic Sci. Rev.* **16**: 37–62.

Wilson, M.R., Polanskey, D., Butler, J.M., DiZinno, J.A., Replogle, J. and Budowle, B. (1995) Extraction, PCR amplification and sequencing of mitochondrial DNA from human hair shafts. *BioTechniques* **18**: 662–669.

Wolf, A., Caliebe, A., Junge, O. and Krawczak, M. (2005) Forensic interpretation of Y-chromosomal DNA mixtures. *Forensic Sci. Int.* **152**: 209–213.

Woolley, A.T., Sensabaugh, G.F. and Mathies, R.A. (1997) High-speed DNA genotyping using microfabricated capillary array electrophoresis chips. *Anal. Chem.* **69**: 2181–2186.

YCC. (2002) A nomenclature system for the tree of human Y chromosomal binary haplogroups. *Genome Res.* **12**: 339–348.

Zamir, A., Springer, E. and Glattstein, B. (2000) Fingerprints and DNA: STR typing of DNA extracted from adhesive tape after processing for fingerprints. *J. Forensic Sci.* **45**: 687–688.

2

Basic tools and techniques in molecular biology

Ralph Rapley and David Whitehouse

2.1 Introduction

The purpose of this chapter is to provide a perspective on basic molecular biology techniques that are of special relevance to forensic genetics. Genetic polymorphisms are the most valuable tools for human identification and for determining genetic relationships and have consequently become a mainstay for forensic science. Throughout the history of forensic genetics, genetic polymorphisms have been studied at various levels from the cellular and serological, such as the determination of blood groups and HLA types, gene product analysis, such as the red cell isozyme and serum protein polymorphisms, to the direct examination of nuclear and mitochondrial DNA. The last has afforded a variety of polymorphic systems, notably fragment length polymorphisms, due to variable number tandem repeats (VNTRs) and single nucleotide polymorphisms (SNPs). DNA analysis offers exquisite resolution compared with the former cellular and gene product analysis approaches to characterizing genetic variation; for the first time it has enabled a direct view of the entire genome. It is the techniques and procedures for working with and analysing segments of DNA that are the subject matter of this chapter.

2.2 Isolation and separation of nucleic acids

Isolation of DNA

The use of DNA for forensic analysis or manipulation usually requires that it is isolated and purified to a certain extent (see Chapter 3). It should be noted that the level of purity of template DNA to be amplified by the polymerase chain reaction is frequently far less critical than the knowledge that the sample to be analysed is uncontaminated and exclusively contains only material from the stated source. Figure 2.1 illustrates a general scheme for DNA extraction.

DNA is recovered from cells by the gentlest possible method of cell rupture to prevent the DNA from fragmenting by mechanical shearing. This is usually in the presence of EDTA, which chelates the Mg^{2+} ions needed for enzymes that degrade DNA, termed DNases. Ideally the cell membrane should be solubilized using detergent, and cell walls, if present, should be digested enzymatically (e.g. lysozyme treatment of bacteria). If physical disruption is necessary, it should be kept to a minimum, and should involve cutting or squashing of cells rather than the use of shear forces. Cell disruption should be performed at 4°C, using disposable plastics where possible; all glassware and solutions are autoclaved to destroy DNase activity (Cseke *et al.*, 2004). Techniques such as laser microdissection are being investigated as exciting new tools for the recovery of genetic material from crime scenes (see Chapter 10). After release of nucleic acids from the cells, RNA can be removed by treatment with ribonuclease (RNase), which

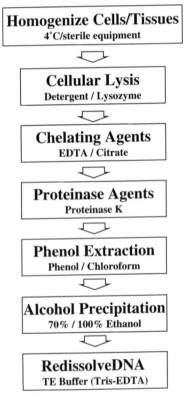

Figure 2.1   General steps involved in extracting DNA from cells or tissues

is usually heat treated to inactivate any DNase contaminants (Cseke *et al.*, 2004). The other major contaminant, protein, is removed by shaking the solution gently with water-saturated phenol, or with a phenol/chloroform mixture, either of which will denature proteins but not nucleic acids. Centrifugation of the emulsion formed by this mixing produces a lower, organic phase, separated from the upper, aqueous phase by an interface of denatured protein. The aqueous solution is recovered and deproteinized repeatedly, until no more material is seen at the interface. Finally, the deproteinized DNA preparation is mixed with sodium acetate and absolute ethanol, and the DNA is allowed to precipitate out of solution in a freezer. After centrifugation, the DNA pellet, which can be washed in 70% ethanol to remove excess salt, is redissolved in a buffer containing EDTA to inactivate any DNases present. This solution can be stored at 4°C for at least a month.

DNA solutions can be stored frozen for prolonged periods, although repeated freezing and thawing tends to damage long DNA molecules. The procedure described above is suitable for total cellular DNA, mitochondrial DNA and DNA from microorganisms and viruses. However, in the case of the extraction of DNA from difficult sources such as hair shafts, a more vigorous preparatory step is required, such as disrupting the starting material in a tissue grinder. The integrity of the DNA can be checked by agarose gel electrophoresis and the concentration of the DNA can be determined spectrophotometrically.

Contaminants may also be identified by scanning UV spectrophotometry from 200 nm to 300 nm. A ratio of 260 : 280 nm of approximately 1.8 indicates that the sample is free of protein contamination, which absorbs strongly at 280 nm.

Automated and kit-based extraction of nucleic acids

Automation and kit-based manipulations in molecular biology are steadily increasing, and the extraction of nucleic acids by these means for forensic analysis is no exception (see Chapter 3). There are many commercially available kits for nucleic acid extraction. Although many rely on the methods described here, their advantage lies in the fact that the reagents are standardized and quality-control-tested, providing a high degree of reliability. Essentially the same reagents for nucleic acid extraction may be used in a format that allows reliable and automated extraction. This is of particular use where a large number of DNA extractions are required (Montpetit *et al.*, 2005).

2.3 Automated analysis of nucleic acid fragments

Gel electrophoresis remains the established method for the separation and analysis of nucleic acids. Indeed a number of automated systems using precast gels are available that are gaining popularity. This is especially useful in situations where a large number of samples or high-throughput analysis is required. In addition, new technologies such as Agilents' Lab-on-a-chip have been developed that obviate the need to prepare electrophoretic gels. These systems employ microfluidic circuits where a small cassette unit that contains interconnected micro-reservoirs is used. The sample is applied in one area and driven through microchannels under computer-controlled electrophoresis. The channels lead to reservoirs allowing, for example, incubation with other reagents such as dyes for a specified time. Electrophoretic separation is thus carried out in a microscale format. The small sample size minimizes sample and reagent consumption, and as such is useful for DNA and RNA sample analysis. In addition the units, being computer controlled, allow data to be captured within a very short timescale (see He *et al.*, 2001). Alternative methods of analysis, including denaturing high-performance liquid chromatography-based approaches, have gained in popularity, especially for mutation analysis (Underhill *et al.*, 2001). Mass spectrometry is also becoming increasingly used for nucleic acid analysis (Oberacher *et al.*, 2006).

2.4   Molecular biology and bioinformatics

Databases and basic bioinformatics

Bioinformatics has become a vital resource for applied forensic molecular biology and is a key component of the routine detection and identification of short tandem repeat (STR) profiles in forensic casework (see Chapter 11). The National DNA Database (NDNAD) established in 1995 was the first forensic science database. It contains STR profiles from subjects in the UK. Samples are normally taken from mouth swabs, though less frequently blood samples are taken. The NDNAD also contains information on samples from volunteers and crime scenes (Parliamentary Office of Science and Technology, 2006). Many countries now maintain their own forensic DNA databases. For example, in the USA the FBI has developed the Combined DNA Index System (CODIS).

The emergence of nucleic acid and the accompanying bioinformatics tools has been driven principally by the Human Genome Project with its need to store, analyse and manipulate vast numbers of DNA sequences. There are now a huge number of sequences stored in genetic databases from a variety of other organisms. The largest of the sequence databases include GenBank at the National Institutes of Health (NIH) in the USA, EMBL at the European Bioinformatics Institute (EBI) at Cambridge, UK and the DNA database of Japan (DDBJ) at Mishima in Japan. All the genome databases are accessible to the public via the Internet.

2.5   The polymerase chain reaction (PCR)

Basic concept of the PCR

The polymerase chain reaction or PCR is currently the mainstay of forensic molecular biology. One of the reasons for the wide adoption of the PCR globally is the elegant simplicity of the reaction and relative ease of the practical manipulation steps. Indeed, combined with the relevant bioinformatics resources for its design and for determination of the required experimental conditions, it provides a rapid means for DNA identification and analysis. It has opened up the investigation of cellular and molecular processes to those outside the field of molecular biology (Altshuler, 2006).

The PCR is used to amplify a precise fragment of DNA from a complex mixture of starting material, usually termed the template DNA, and in many cases requires little DNA purification. It does require the knowledge of some DNA sequence information, which flanks the fragment of DNA to be amplified (target DNA). From this information two oligonucleotide primers may be chemically synthesized, each complementary to a stretch of DNA to the 3' side of the target DNA, one oligonucleotide for each of the two DNA strands (Figure 2.2). It may be thought of as a technique analogous to the DNA replication process that takes place in cells since the outcome is the same: the generation of new complementary DNA stretches based upon the existing ones. It is also a technique that has replaced, in many cases, the traditional DNA cloning methods since it fulfils the same function – the production of large amounts of DNA from limited starting material – however this is achieved in a fraction of the time needed to clone a DNA fragment. Although not without its drawbacks, the PCR is a remarkable development that is changing the approach of many scientists to the analysis of nucleic acids and continues to have a profound impact on core biosciences and biotechnology.

Stages in the PCR

The PCR consists of three defined sets of times and temperatures, termed steps: (i) denaturation, (ii) annealing and (iii) extension. Each of these steps is repeated 30–40 times, termed cycles (Figure 2.3). In the first cycle the double-stranded template DNA is (i) denatured by heating the reaction to above 90°C. Within the complex DNA the region to be specifically amplified (target) is made accessible. The temperature is then cooled to between 40 and 60°C. The precise temperature is critical and each PCR system has to be defined and optimized. One useful technique for optimization is Touchdown PCR where a programmable cycler is used to incrementally decrease the annealing temperature until the optimum is derived. Reactions that are not optimized may give rise to other

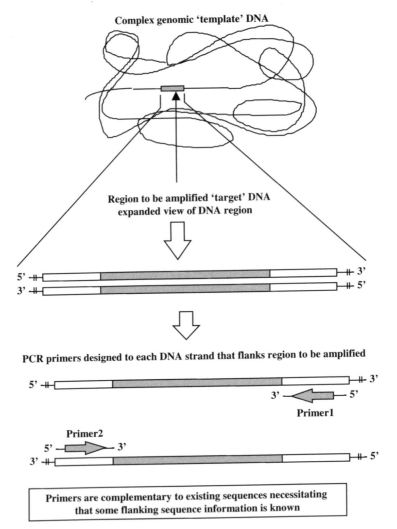

Figure 2.2   The location of PCR primers: PCR primers designed to sequences adjacent to the region to be amplified, allowing a region of DNA (eg. a gene) to be amplified from a complex starting material of genomic template DNA

DNA products in addition to the specific target or may not produce any amplified products at all. The annealing step allows the hybridization of the two oligonucleotide primers, which are present in excess, to bind to their complementary sites that flank the target DNA. The annealed oligonucleotides act as primers for DNA synthesis, since they provide a free 3′ hydroxyl group for DNA polymerase. The DNA synthesis step is termed extension and is carried out by a thermostable DNA polymerase, most commonly *Taq* DNA polymerase.

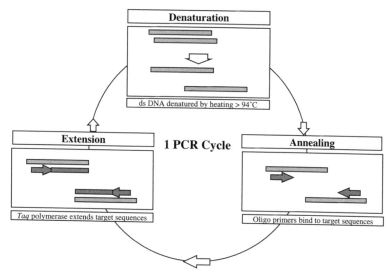

Figure 2.3 Simplified scheme of one PCR cycle that involves denaturation, annealing and extension

DNA synthesis proceeds from both of the primers until the new strands have been extended along and beyond the target DNA to be amplified. It is important to note that, since the new strands extend beyond the target DNA, they will contain a region near their 3′ ends that is complementary to the other primer. Thus, if another round of DNA synthesis is allowed to take place, not only the original strands will be used as templates but also the new strands. Most interestingly, the products obtained from the new strands will have a precise length, delimited exactly by the two regions complementary to the primers. As the system is taken through successive cycles of denaturation, annealing and extension, all the new strands will act as templates and so there will be an exponential increase in the amount of DNA produced. The net effect is to selectively amplify the target DNA and the primer regions flanking it.

One problem with early PCR reactions was that the temperature needed to denature the DNA also denatured the DNA polymerase. However, the availability of a thermostable DNA polymerase enzyme isolated from the thermophilic bacterium *Thermus aquaticus* found in hot springs provided the means to automate the reaction. *Taq* DNA polymerase has a temperature optimum of 72°C and survives prolonged exposure to temperatures as high as 96°C, so it is still active after each of the denaturation steps. The widespread utility of the technique is also due to the ability to automate the reaction and, as such, many thermal cyclers have been produced in which it is possible to program in the temperatures and times for a particular PCR reaction.

Polymerase chain reaction primer design and bioinformatics

The specificity of the PCR lies in the design of the two oligonucleotide primers. These have to be complementary to sequences flanking the target DNA but must not be self-complementary or bind each other to form dimers since both prevent DNA amplification. They also have to be matched in their GC content and have similar annealing temperatures. The increasing use of bioinformatics resources such as Oligo, Generunner and Primer Design Assistant (Chen *et al.*, 2003) in the design of primers makes the design and selection of reaction conditions much more straightforward. These resources allow the sequences to be amplified, the primer length, product size, GC content, etc. to be input and, following analysis, provide a choice of matched primer sequences. Indeed the initial selection and design of primers without the aid of bioinformatics would now be unnecessarily time-consuming.

Polymerase chain reaction amplification templates

DNA from a variety of sources may be used as the initial source of amplification templates. It is also a highly sensitive technique and requires only one or two molecules for successful amplification, thus enabling the genetic material of a single cell to be analysed. This property of the PCR has been exploited by forensic scientists through the use of low copy number (LCN) DNA profiling. The sensitivity of the LCN profiling process depends on an increased number of PCR cycles, typically 34. Unlike many manipulation methods used in current molecular biology, the PCR technique is sensitive enough to require very little template preparation. The PCR may also be used to amplify RNA, a process termed reverse transcriptase PCR (RT-PCR) (Gill *et al.*, 2000).

Sensitivity of the PCR

The enormous sensitivity of the PCR system is also one of its main drawbacks since the very large degree of amplification makes the system vulnerable to contamination. Even a trace of contaminating DNA, such as that contained in dust particles, may be amplified to significant levels and may give misleading results. Hence a rigorous sample-handling protocol is essential for casework when carrying out the PCR and dedicated equipment and even laboratories are preferable. All precautions must be taken to prevent previously amplified products (amplicons) from contaminating the PCR.

2.6 Applications of the PCR

Many traditional methods in molecular biology have now been superseded by the PCR and the applications for the technique appear to be unlimited. The success of the PCR process has given impetus to the development of other amplification techniques, which are based on either thermal cycling or non-thermal cycling (isothermal) methods. The most popular alternative to the PCR is termed the ligase chain reaction (LCR). This operates in a similar fashion to the PCR but a thermostable DNA ligase joins sets of primers together that are complementary to the target DNA. Following this a similar exponential amplification reaction takes place, producing amounts of DNA that are similar to the PCR.

Quantitative and real-time PCR

One of the most useful PCR applications is quantitative PCR (Q-PCR) (see Chapter 4). Quantitative PCR is gaining popularity in forensic science mainly because of the rapidity of the method compared to conventional PCR amplification whilst simultaneously providing a lower limit of detection and greater dynamic range. Another advantage is that Q-PCR enables a rigorous analysis of PCR problems as they arise. Early quantitative PCR methods involved the comparison of a standard or control DNA template amplified with separate primers at the same time as the specific target DNA (Higuchi *et al.*, 1993). These types of quantitation rely on the reaction being exponential and so any factors affecting this may also affect the result. Other methods involve the incorporation of a radiolabel through the primers or nucleotides and their subsequent detection following purification of the amplicon. An alternative automated real-time PCR method is the 5′ fluorogenic exonuclease detection system or TaqMan assay (Holland *et al.*, 1991). In its simplest form a DNA binding dye such as SYBR Green is included in the reaction. As amplicons accumulate, SYBR Green binds the double-stranded DNA proportionally. Fluorescence emission of the dye is detected following excitation. The binding of SYBR Green is non-specific, therefore in order to detect specific amplicons an oligonucleotide probe labelled with a fluorescent reporter and quencher molecule at either end is included in the reaction in the place of SYBR Green. When the oligonucleotide probe binds to the target sequence the 5′ exonuclease activity of *Taq* polymerase degrades and releases the reporter from the quencher. A signal is thus generated that increases in direct proportion to the number of starting molecules. Thus a detection system is able to induce and detect fluorescence in real-time as the PCR proceeds. In addition to quantitation, real-time PCR systems may also be used for genotyping and for accurate determination of amplicon melting temperature using curve analysis. This allows accurate amplicon identification and also offers the potential to detect mutations and SNPs (see Chapter 6). Further developments in probe-based PCR systems have also been used and include scorpion probe systems (Solinas *et al.*, 2001), amplifluor and real-time LUX probes.

2.7   Nucleotide sequencing of DNA

Concepts of nucleic acid sequencing

The determination of the order or sequence of bases along a length of DNA is one of the central techniques in molecular biology and has a key role to play in the development of polymorphic systems and analysis of the mitochondrial genome in forensic science (see Chapter 8). The precise usage of codons, information regarding mutations and polymorphisms and the identification of gene regulatory control sequences are also only possible by analysing DNA sequences. Two techniques have been developed for this, one based on an enzymatic method, frequently termed Sanger sequencing after its developer, and a chemical method called Maxam and Gilbert, named for the same reason. At present Sanger sequencing is by far the most popular method and many commercial kits are available for its use. However, there are certain occasions, such as the sequencing of short oligonucleotides, where the Maxam and Gilbert method is more appropriate (Kieleczawa, 2005).

One absolute requirement for Sanger sequencing is that the DNA to be sequenced is in a single-stranded form. Traditionally this demanded that the DNA fragment of interest be inserted and cloned into a specialized bacteriophage vector termed M13, which is naturally single-stranded. Although M13 is still universally used the advent of the PCR has provided the means to not only amplify a region of any genome or complementary DNA but also very quickly to generate the corresponding nucleotide sequence. This has led to an explosion in the accumulation of DNA sequence information and has provided much impetus for gene discovery and genome mapping.

The Sanger method is simple and elegant and mimics in many ways the natural ability of DNA polymerase to extend a growing nucleotide chain based on an existing template. Initially the DNA to be sequenced is allowed to hybridize with an oligonucleotide primer, which is complementary to a sequence adjacent to the 3′ side of DNA within a vector such as M13 or in an amplicon. The oligonucleotide will then act as a primer for the synthesis of a second strand of DNA, catalysed by DNA polymerase. Since the new strand is synthesized from its 5′ end, virtually the first DNA to be made will be complementary to the DNA to be sequenced. One of the deoxyribonucleoside triphosphates (dNTPs) that must be provided for DNA synthesis is radioactively labelled with ^{32}P or ^{35}S, and so the newly synthesized strand will be labelled.

Dideoxynucleotide chain terminators

The reaction mixture is then divided into four aliquots, representing the four dNTPs A, C, G and T. In addition to all of the dNTPs being present in the A tube, an analogue of dATP is added (2′,3′-dideoxyadenosine triphosphate, ddATP) that is similar to A but has no 3′ hydroxyl group and so will terminate the growing

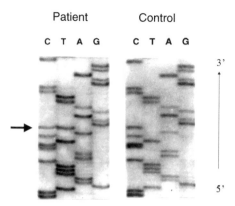

Figure 2.4 Part of Exon 2 of the human PI gene that encodes alpha-1 antitrypsin, illustrating the technique of manual sequencing with ^{35}S and autoradiography. Note that the patient is heterozygous for a C and T where indicated by the arrow

chain because a 5′ to 3′ phosphodiester linkage cannot be formed without a 3′-hydroxyl group. The situation for tube C is identical except that ddCTP is added; similarly the G and T tubes contain ddGTP and ddTTP, respectively.

Since the incorporation of ddNTP rather than dNTP is a random event, the reaction will produce new molecules varying widely in length, but all terminating at the same type of base. Thus four sets of DNA sequence are generated, each terminating at a different type of base, but all having a common 5′ end (the primer). The four labelled and chain-terminated samples are then denatured by heating and loaded next to each other on a polyacrylamide gel for electrophoresis. Electrophoresis is performed at approximately 70°C in the presence of urea, to prevent renaturation of the DNA, since even partial renaturation alters the rates of migration of DNA fragments. Very thin, long gels are used for maximum resolution over a wide range of fragment lengths. After electrophoresis, the positions of radioactive DNA bands on the gel are determined by autoradiography. Since every band in the track from the ddATP sample must contain molecules that terminate at adenine, and those in the ddCTP that terminate at cytosine, etc., it is possible to read the sequence of the newly synthesized strand from the autoradiogram, provided that the gel can resolve differences in length equal to a single nucleotide, hence the ability to detect and characterize point mutations (Figure 2.4). Under ideal conditions, sequences up to about 300 bases in length can be read from one gel.

Direct PCR pyrosequencing

Rapid PCR sequencing has also been made possible by the use of pyrosequencing. This is a sequencing by synthesis whereby a PCR template is hybridized to an oligonucleotide and incubated with DNA polymerase, ATP sulphurylase, luciferase and apyrase. During the reaction the first of the four dNTPs is added and, if incorporated, it releases pyrophosphate (PPi). The ATP sulphurylase converts the PPi to ATP, which drives the luciferase-mediated conversion of luciferin to oxyluciferin in order to generate light. Apyrase degrades the resulting component dNTPs and ATP. This is followed by another round of dNTP addition. A resulting pyrogram provides an output of the sequence. The method provides short reads very quickly and is especially useful for the determination of mutations or SNPs (Ronaghi *et al.*, 1998).

It is also possible to undertake nucleotide sequencing from double-stranded molecules such as plasmid cloning vectors and PCR amplicons directly. The double-stranded DNA must be denatured prior to annealing with primer. In the case of plasmid an alkaline denaturation step is sufficient, however for amplicons this is more problematic and a focus of much research. Unlike plasmids, amplicons are short and re-anneal rapidly, thereby preventing the re-annealing process or biasing the amplification towards one strand by using a primer ratio of 100 : 1 to overcome this problem to a certain extent. Denaturants such as formamide or dimethylsulphoxide (DMSO) have also been used with some success in preventing the re-annealing of PCR strands following their separation.

It is possible to physically separate and retain one PCR strand by incorporating a molecule such as biotin into one of the primers. Following PCR one strand with an affinity molecule may be removed by affinity chromatography with strepavidin, leaving the complementary PCR strand. This affinity purification provides single-stranded DNA derived from the PCR amplicon and although it is somewhat time-consuming it does provide high-quality single-stranded DNA for sequencing.

Polymerase chain reaction cycle sequencing

One of the most useful methods of sequencing PCR amplicons is termed PCR cycle sequencing. This is not strictly a PCR since it involves linear amplification with a single primer. Approximately 20 cycles of denaturation, annealing and extension take place. Radiolabelled or fluorescent-labelled dideoxynucleotides are then introduced in the final stages of the reaction to generate the chain-terminated extension products. Automated direct PCR sequencing is increasingly being refined, allowing greater lengths of DNA to be analysed in one sequencing run, and provides a very rapid means of analysing DNA sequences (Dugan *et al.*, 2002).

Automated DNA sequencing

Advances in fluorescent dye terminator and labelling chemistry have led to the development of high-throughput automated sequencing techniques. Essentially most systems involve the use of dideoxynucleotides labelled with different fluorochromes. The advantage of this modification is that since a different label is incorporated with each ddNTP it is unnecessary to perform four separate reactions. Therefore the four chain-terminated products are run on the same track of a denaturing electrophoresis gel. Each product with their base-specific dye is excited by a laser and the dye then emits light at its characteristic wavelength. A diffraction grating separates the emissions, which are detected by a charge-coupled device (CCD), and the sequence is interpreted by a computer. The advantages of these techniques include real-time detection of the sequence. In addition, the lengths of sequence that may be analysed are in excess of 500 bp. Capillary electrophoresis is increasingly being used for the detection of sequencing products (Plate 2.1). This is where liquid polymers in thin capillary tubes are used, obviating the need to pour sequencing gels and requiring little manual operation. This substantially reduces the electrophoresis run times and allows high throughput to be achieved. A number of large-scale sequence facilities are now fully automated using 96-well microtitre-based formats. The derived sequences can be downloaded automatically to databases and manipulated using a variety of bioinformatics resources. Developments in the technology of DNA sequencing have made whole genome sequencing projects a realistic proposition within achievable time-scales, and a number of these have been or are nearing completion (Kline *et al.*, 2005).

Maxam and Gilbert sequencing

Sanger sequencing is by far the most popular technique for DNA sequencing, however an alternative technique developed at the same time may also be used. The chemical cleavage method of DNA sequencing developed by Maxam and Gilbert is often used for sequencing small fragments of DNA such as oligonucleotides, where Sanger sequencing is problematic. A radioactive label is added to either the 3' or the 5' ends of a double-stranded DNA. The strands are then separated by electrophoresis under denaturing conditions, and analysed separately. DNA labelled at one end is divided into four aliquots and each is treated with chemicals that act on specific bases by methylation or removal of the base. Conditions are chosen so that, on average, each molecule is modified at only one position along its length; every base in the DNA strand has an equal chance of being modified. Following the modification reactions, the separate samples are cleaved by piperidine, which breaks phosphodiester bonds exclusively at the 5' side of nucleotides whose base has been modified. The result is similar to that produced by the Sanger method, since each sample now contains radioactively labelled molecules of various lengths, all with one end in common (the labelled end) and with the other end cut at the same type of base. Analysis of the reaction products by electrophoresis is as described for the Sanger method.

2.8   Conclusion

The impact of molecular biology on forensic science has been massive and far-reaching. The combined information content of molecular polymorphisms has literally revolutionized the aim of the scientists to the extent that exclusion probabilities have given way to positive identification of individuals matched with evidential material (see Chapter 12). Bioinformatics and greater emphasis on mapping complex trait genes could lead to the identification of DNA markers for many common characteristics, enabling crime detection at the levels of the genotype and phenotype simultaneously. The future is also likely to witness the widespread introduction of genotyping microchips for both nucleic acids and proteins. Proteomics (see Chapter 13) offers the significant potential of utilizing gene products for the advancement of forensic analysis.

2.9   References

Altshuler, M.L. (2006) *PCR Troubleshooting: The Essential Guide*, Caister Academic Press, Wymondham, UK.

Chen, S.H., Lin, C.Y., Cho, C.S., Lo, C.Z. and Hsiung, C.A. (2003) Primer Design Assistant (PDA): a web-based primer design tool. *Nucleic Acids Res.* **31**: 3751–3754.

Cseke, L.J., Kaufman, P.B., Podila, G.K. and Tsai, C.J. (2004) *Handbook of Molecular and Cellular Methods in Biology and Medicine* (2nd edn), CRC Press, Boca Raton, FL.

Dugan, K.A., Lawrence, H.S., Hares, D.R., Fisher, C.L. and Budowle, B. (2002) An improved method for post-PCR purification for mtDNA sequence analysis. *J. Forensic Sci.* **47**: 811–818.

Gill, P., Whitaker, J., Flaxman, C., Brown, N. and Buckelton, J. (2000) An investigation of the rigor of interpretation rules for STRs derived from less than 100 pg of DNA. *Forensic Sci. Int.* **112**: 17–40.

Higuchi, R., Fockler, C., Dollinger, G. and Watson, R. (1993) Kinetic PCR: Real-time monitoring of DNA amplification reactions. *Biotechnology* **11**: 1026–1030.

He, Y., Zhang, Y.H. and Yeung, E.S. (2001) Capillary-based fully integrated and automated system for nanoliter polymerase chain reaction analysis directly from cheek cells. *J. Chromatogr. A* **924**: 271–284.

Holland, P.M., Abramson, R.D., Watson, R. and Gelfand, H. (1991) Detection of specific polymerase chain reaction product by utilizing the 5′-3′exonuclease activity of Thermus Aquaticus DNA polymerase *Proc. Natl. Acad. Sci. USA* **88**: 7276–7280.

Kieleczawa, J. (2005) *DNA Sequencing*, Jones and Bartlett Publishers, Sudbury, USA.

Kline, M.C., Vallone, P.M., Redman, J.W., Duewer, D.L., Calloway, C.D. and Butler, J.M. (2005) Mitochondrial DNA typing screens with control region and coding region SNPs. *J. Forensic Sci.* **50**: 377–385.

Montpetit, S.A., Fitch, I.T. and O'Donnell, P.T. (2005) A simple automated instrument for DNA extraction in forensic casework. *J. Forensic Sci.* **50**: 555–563.

Oberacher, H., Niederstatter, H. and Parson, W. (2006) Liquid chromatography-electrospray ionization mass spectrometry for simultaneous detection of mtDNA length and nucleotide polymorphisms. *Int. J. Legal Med.* [Epub ahead of print].

Parliamentary Office of Science and Technology (2006) Feb 2006, Number 2008.

Ronaghi, M., Uhlen, M. and Nyren, P. (1998) A sequencing method based on real-time pyrophosphate. *Science* **281**: 363–365.

Solinas, A., Brown, L.J., McKeen, C., Mellor, J.M., Nicol, J., Thelwell, N. and Brown, T. (2001) Duplex Scorpion primers in SNP analysis and FRET applications. *Nucleic Acids Res.* **29**: E96.

Underhill, P.A., Passarino, G., Lin, A.A., Shen, P., Mirazon Lahr, M., Foley, R.A., Oefner, P.J. and Cavalli-Sforza, L.L. (2001) The phylogeography of Y chromosome binary haplotypes and the origins of modern human populations. *Ann. Hum. Genet.* **65**: 43–62.

3

Automated DNA extraction techniques for forensic analysis

Marion Nagy

3.1 Introduction

Many different methods are used to extract DNA from the wide range of specimens commonly found at crime scenes. Techniques range from simple alkaline lysis followed by neutralization (Klintschar and Neuhuber, 2000), to the well-known salting-out method (Miller *et al.*, 1988) that is used in cases of higher cell concentrations, to a simple closed-tube method utilizing a thermostable proteinase (Moss *et al.*, 2003). Only a few of these methods are suitable for automation, and many of the steps involved are associated with a high risk of contamination. Many methods also require centrifugation and solvent extraction steps (Sambrook *et al.*, 1989; Walsh *et al.*, 1991) and thus are also not easily adapted for automation. Most of the methods used for DNA extraction can meet only one of the several standards for an optimal DNA extraction process: high DNA yield, rapidity of the method (McHale *et al.*, 1991), high throughput and high DNA quality (Akane *et al.*, 1994; Klintschar and Neuhuber, 2000). Thus, for a long time the method of choice for many forensic samples was the traditional but hazardous use of phenol–chloroform extraction, which had to be performed under stringent safety measures (Butler, 2005). The end-product of the extraction sometimes was not sufficiently pure and still needed to undergo further purification steps using membranes or columns, so this method is not ideal in several respects. In order to assess the utility of other methodologies for DNA extraction, we first examine the principal steps in isolating DNA from biological material.

Molecular Forensics.   Edited by Ralph Rapley and David Whitehouse
Copyright 2007 by John Wiley & Sons, Ltd.

3.2   Principal steps of DNA extraction

Cell lysis

The first step in any DNA extraction method is to break the cells open in order to access the DNA within. Although DNA may be isolated by 'boiling' cells (Starnbach *et al.*, 1989), this rather crude means of disrupting the cell does not produce DNA that is always of sufficient quality and purity to be used in down-stream analytical techniques such as polymerase chain reaction (PCR) amplifica-tion. DNA isolated by simple boiling generally fails as a substrate for further analysis because it has not been sufficiently separated from structural elements and DNA-binding proteins, and these impurities compromise downstream pro-cedures. In order for DNA to be released cleanly, the phospholipid cell mem-branes and nuclear membranes have to be disrupted in a process called lysis, which uses a detergent solution (lysis buffer), often containing the detergent sodium dodecyl sulphate (SDS), which disrupts lipids and thus disrupts mem-brane integrity. Lysis buffer also contains a pH-buffering agent to maintain the pH of the solution so that the DNA stays stable: DNA is negatively charged due to the phosphate groups on its structural backbone, and its solubility is charge-dependent and thus pH-dependent. Proteinases, which are enzymes that digest proteins, are generally added to lysis buffer in order to remove proteins bound to the DNA and to destroy cellular enzymes that would otherwise digest DNA upon cell lysis. The lysis procedure sometimes calls for the use of heat and agitation in order to speed up the enzymatic reactions and the lipid solubilization.

DNA extraction: purification and efficient removal of PCR inhibitors

Cell or tissue samples may contain elements that inhibit the DNA extraction process at any of the various steps involved in DNA isolation. These inhibitors may interfere at any step of the process, but are generally problematic in three areas:

1. Interference with the cell lysis, the first step in DNA preparation.

2. Interference by degrading nucleic acids or by otherwise preventing their isola-tion after lysis is complete.

3. Inhibition of polymerase activity during the PCR amplification of target DNA after successful purification.

We focus here on the third type of inhibition: interference with PCR. After the initial isolation of DNA, it must be separated from the other cellular components that remain after the lysis procedure. This is often followed by further washing steps, which function to remove any remaining substances that could inhibit amplification of the DNA by PCR and its subsequent analysis. A wide range of PCR inhibitors have been reported (see Wilson, 1997, for review). Common inhibitors include various body fluid components (e.g. haemoglobin, melanin, urea) as well as chemical reagents that are frequently used in clinical and forensic science laboratories (e.g. heparin, formalin, Ca^{2+}). Inhibitors also include microorganism populations, which are frequently an overrepresentation of bacterial cells or food constituents found at the scene. Similarly, environmental compounds at the crime scene or in the forensic laboratory can also act as PCR inhibitors. These include organic and phenolic compounds, glycogen, polysaccharides, humic acids, fats and laboratory items such as pollen, glove powder and plasticware residue. Carry-over of compounds such as those used in cell lysis (e.g. proteolytic enzymes or denaturants) and phenolic compounds from DNA purification procedures can also be problematic. Many of these PCR-inhibitory compounds, such as polysaccharides, urea, humic acids, haemoglobin, melanin in hair samples or indigo dyes from denim, exhibit a solubility similar to that of DNA, therefore they are not completely removed during classical extraction protocols such as detergent and phenol–chloroform extraction, and persist as contaminants in the final DNA preparation. Several methods have been developed to remove these contaminants, including glass bead extraction, size-exclusion chromatography, spin column separation, agarose-embedded DNA preparation or immunomagnetic separation (Wilson, 1997; Moreira, 1998).

3.3 DNA extraction techniques

There are, in general, three primary techniques used in forensic DNA laboratories: the phenol–chloroform extraction method, Chelex extraction and magnetic affinity solid-phase extraction. The principles of all three methods are shown in Figure 3.1 and are described in the following sections.

Standard phenol–chloroform extraction

The standard phenol–chloroform extraction protocol in use today (Figure 3.1a) is described in Sambrook *et al.* (1989). Samples are incubated with enzymatic lysis buffer (e.g. 10 mM Tris·HCl pH 7.4, 400 mM NaCl, 2 mM Na_2EDTA pH 8.1, 1% SDS and 667 µg/ml proteinase K) overnight at 37°C or for 2 hours at 56°C to partially digest cellular proteins. The resultant liquid-phase cell lysate is treated with equilibrated phenol, and the aqueous and organic phases are mixed thoroughly and then separated by centrifugation. The DNA remains in the aqueous phase while the cellular proteins are extracted into the organic

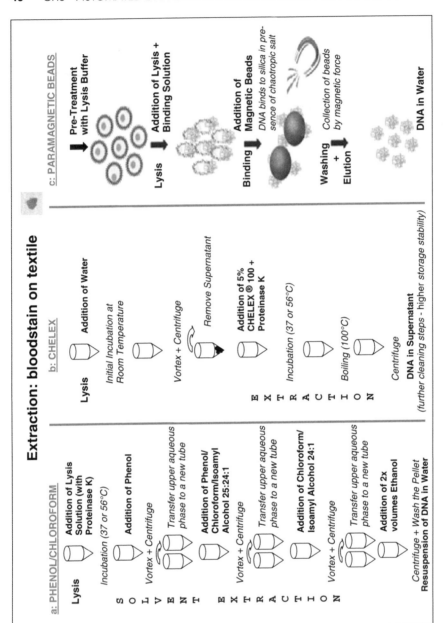

Figure 3.1   Principles of extraction techniques commonly used in forensic DNA laboratories. Part (c) reproduced with permission from QIAGEN Products

phase and discarded. The aqueous supernatant is transferred to a new tube and an equilibrated mixture of phenol–chloroform–isoamyl alcohol in the ratio of 25:24:1 is added to ensure complete removal of the proteins. After agitation, the mixture is centrifuged and the resultant aqueous supernatant is transferred again to a new tube and a chloroform–isoamyl alcohol mixture in the ratio of 24:1 is added and followed by further mixing and centrifugation in order to ensure the complete removal of phenol. The aqueous supernatant is collected, and an acetate salt and 2 volumes of 96% ice-cold ethanol are added to precipitate the DNA via shielding of the negative charges on DNA, allowing it to aggregate and precipitate. Samples are incubated at −20°C to increase the efficiency of precipitation, and are then centrifuged to collect the precipitated DNA. Ethanol is decanted and the DNA pellet is washed with 70% ethanol to remove the salt, then vacuum- or air-dried and resuspended in either sterile double-distilled water or low-salt buffer. Some protocols involve an additional purification and concentration step with spin columns, which remove inhibitors with solubility characteristics similar to DNA, such as heme (Akane *et al.*, 1994) or indigo dyes.

Phenol–chloroform extraction works very well for the recovery of double-stranded high-molecular-weight DNA. However, the method is time-consuming, involves the use of hazardous chemicals and requires multiple tube transfers as well as a final precipitation step, potentially increasing the risk of contamination and/or sample mix-ups. Nevertheless, the method works very well for extraction of DNA from nearly all of the common types of forensic samples, and is still used today as a last resort for DNA extraction from problematic samples because it produces relatively large yields of high-quality DNA, and the excellent DNA purity allows it to remain stable in long storage. In this respect, phenol–chloroform extraction remains the gold standard by which new methods are judged.

Chelex® 100 extraction

Chelex® 100 is a medium used for the simple extraction of DNA (Figure 3.1b) for subsequent use in PCR-based typing. It is a styrene divinylbenzene copolymer containing paired iminodiacetate ions, which act as chelating groups in the binding of polyvalent metal ions such as magnesium or calcium (Instruction Manual, Chelex® 100, Bio-Rad Laboratories, Hercules, CA). Chelex therefore binds bivalent ions such as Ca^{2+} and Mg^{2+} and deactivates unwanted enzymes such as DNases. Chelation of these cations may also result in the 'deactivation', via structural changes, of proteins that make up the cellular architecture, leading to the destabilization of the whole cell and essentially resulting in cellular lysis. Addition of proteinase K, a Ca^{2+}-independent enzyme, breaks down the deactivated enzymes and proteins. Proteinase K is then itself deactivated by a subsequent boiling step. This method was introduced into forensic laboratories on the basis of a protocol by Walsh *et al.* (1991): biological samples are added to a 5% Chelex® 100 resin, are boiled for several minutes and are then centrifuged to remove the Chelex resin, leaving the DNA in the supernatant. The single-stranded DNA can be used directly in PCR. Better results are often obtained if an initial incubation step is used in order to remove the 'forensic stain' from the carrier. Besides, several protocols call for the additional incubation of 100 ng of proteinase K with the Chelex/stain mixture for two hours at 56°C or overnight at 37°C. The Chelex® 100-based extraction has a definite advantage over other methods in that it is very fast and can also be carried out in a single tube without any transfer steps, which substantially reduces the possibility of contamination. However, DNA extracted by Chelex® 100 requires further purification to reach the level of DNA quality obtained from a standard phenol–chloroform extraction.

Magnetic affinity solid-phase extraction

Solid-phase DNA extractions have been in use for many years. In the advanced case of magnetic affinity solid-phase extraction, efficient DNA isolation relies on the binding of DNA to a silica surface on paramagnetic beads in the presence of chaotropic solutions (Figure 3.1c). DNA isolation can therefore be performed in a single tube by adding and removing solutions such that contaminants are washed away while the DNA remains bound to the beads (until its eventual elution). Solid-phase extraction approaches are marketed by Qiagen GmbH, Hilden, Germany (MagAttract® DNA chemistry) and Promega Corporation, Madison, USA (DNA IQ™ system). This extraction method is made possible by the tendency of DNA to bind to silica (glass) in the presence of chaotropic salts such as sodium iodide, guanidinium thiocyanate (GTC) or guanidinium hydrochloride. Cells are first lysed in a lysis and binding solution so that DNA is released. DNases are denatured and inactivated by the presence of the chaotropic salts, and magnetic beads are then mixed with the sample to allow DNA to bind. DNA requires high salt conditions to bind to the silica surface on the beads, but this DNA binding is reversible at pH <7.5, so when the washing steps are complete the DNA may be eluted by water or by a low-salt buffer. After binding, the magnetic beads containing the immobilized DNA are collected by simply applying a magnetic force. The soluble portion containing the unbound components (proteins, cell debris' etc.) is then removed and discarded. The magnetic particles with the attached DNA are then resuspended in a series of wash solutions in order to obtain highly purified DNA: a solution of chaotropic salts removes residual non-bound matter, ethanol removes residual chaotropic salts and a short rinse with water removes ethanol. The DNA is finally eluted from the magnetic beads by the addition of either water or low-salt buffer and is ready for use in downstream applications. Magnetic bead-based DNA isolation has several advantages over both the phenol–chloroform and Chelex® 100 methods, including:

- elimination of traditional solvent extraction, thus avoiding the use of harmful organic solvents,

- elimination of precipitation, centrifugation and pellet-drying steps,

- rapid purification,

- removal of nearly all contaminants that could interfere with subsequent PCR,

- production of high-quality single-stranded DNA,

- scalable and reproducible extraction,

- suitability for adaptation for high-throughput extraction.

3.4   Modified techniques for DNA extraction from challenging forensic samples

Magnetic bead-based purification is currently the technique best suited for DNA extraction from the majority of forensic samples. However, some sample types, such as sperm and skeletal remains, pose special challenges. For these more recalcitrant samples, extraction techniques must be modified in order to successfully extract usable DNA, as we discuss in the following sections.

Sperm extraction – differential extraction

A special differential lysis treatment for forensic samples from sexual assault cases can separate epithelial cells from sperm cells; the most commonly used method was first described by Gill *et al.* (1985). Sperm nuclei are resistant to lysis by the usual SDS/proteinase K cell method, but can be lysed in a solution containing proteinase K plus the reductant dithiothreitol (DTT, e.g. 20 µl of 0.1 M DTT is added to 500 µl of lysis buffer), which breaks down the protein disulphide bridges present in sperm nuclear membranes. This method, known as differential lysis, is often used in forensic laboratories to separate sperm nuclei from vaginal cellular debris in samples obtained from semen-contaminated vaginal swabs, thus enabling the identification of cells specific to the male suspect in a sample of predominantly female cells. However, sperm cells would be absent in a number of situations: some perpetrators of sexual assaults could have a vasectomy or could be azoospermic (lacking any viable sperm), or may have a condition of either retrograde ejaculation (an emission of semen back into the bladder) or anejaculation (complete failure to emit semen). The code and spirit of criminal law states that a violation exists when a sexual act is enforced that highly humiliates the victim, especially when this act entails forced

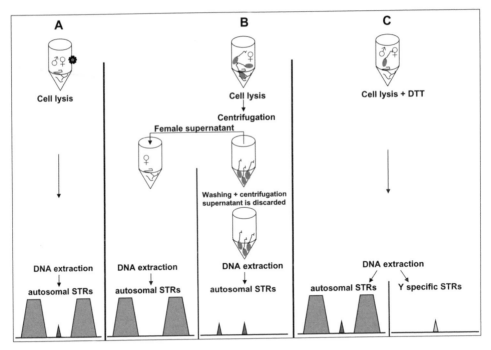

Figure 3.2 Scheme of Selective extraction versus Y-chromosomal analysis: (a) extraction of male and female epithelial cells after cell lysis and detection of autosomal STR profiles; (b) differential extraction to separate male sperm cells from female epithelial cells and detection of autosomal STR profiles; (c) extraction of male and female epithelial cells as well as male sperm cells after cell lysis with DTT and direct analysis of the female (autosomal) and male (Y) STR profile

penetration into the body (§177 StGB, Germany: Gesetzestext mit Rechtssprechung via dejure.org). In the special cases described above, a forensic examination would not reveal sperm cells but would reveal male-specific epithelial cells, providing evidence of a violation. Male DNA profiles can also be picked out of a background of predominantly female cells by analysing for the presence of Y chromosome-specific markers. This Y-specific short tandem repeat (STR) haplotype analysis, which is now as sensitive as the autosomal analysis, has superseded and replaced differential extraction in many forensic laboratories (Prinz *et al.*, 2001; Parson *et al.*, 2003; Nagy *et al.*, 2005). As an indication for sperm cells, DTT is generally added to the cell lysis buffer. Figure 3.2 illustrates the principles behind both methods.

Extraction of DNA from bone

There are several protocols that describe different methods for the extraction of DNA from bones. All methods include freezing of the bone in liquid nitrogen and pulverization with a laboratory mill or a dentist's drill, often followed by decalcification to remove the bone matrix. Such methods were applied in the high-throughput bone-fragment-based DNA identification that was necessary after mass disasters such as the World Trade Center attack (Holland *et al.*, 2003) or the Southeastern Asia tsunami disaster (Steinlechner *et al.*, 2005). The methods differ from each other by what is used for the decalcification step. Prado *et al.* (1997) used only a short decalcification step that entailed overnight incubation of the bone with 0.5 M EDTA, while Holland *et al.* (1993) repeated this EDTA step three times. Höss and Pääbo (1993) extracted the DNA from bone meal without a decalcification step, and instead used guanidinium thiocyanate and the detergent Triton X-100 for cell lysis. However, in this case DNA could be extracted only from those cells that were exposed after bone pulverization. In a compact bone structure, the density of the surrounding cells provides protection against bacterial degradation for a time, but the density also makes it necessary to provide a thorough decalcification step to release the cells from the matrix (Holland *et al.*, 1993). This decalcification step was noted to be especially important in cases of DNA extraction from older bones (Nagy *et al.*, 2005).

However, the need for the initial steps of manual extraction (freezing, grinding) in all of these methods makes it difficult to apply automation to the extraction of DNA from bone. When automation is required as part of a high-throughput effort, the only means by which it can be incorporated into part of the procedure is as follows: manual pre-extraction treatment is followed by thorough decalcification with EDTA. Bulk material is separated by precise centrifugation, and cellular material is lysed, after which a sensitive automated DNA extraction procedure may be employed.

3.5 Automation of DNA extraction

Automated procedures for forensic DNA analyses are a key for high-throughput sample preparation as well as for avoidance of errors during routine sample preparation and for reproducible processing and improved sample tracking. The most important stage in PCR-based forensic analysis is DNA isolation, and both high yields and high purity are vital components of a successful analysis. A wide variety of high-quality automated instruments, each with its own unique benefits and features, are currently available. An overview of the various instruments and the isolation chemistry that each uses is given in Table 3.1. The National Genetics Reference Laboratory (Mattocks, November 2004, Evaluation Report, NGRL, Wessex, UK: http://www.ngrl.co.uk/Wessex/extraction.htm) evaluated eight automated extraction systems covering both dedicated instruments with their associated chemistries (so-called 'integrated systems') as well as kit-based chemistries that may be suitable for automation when adapted to standard liquid-handling robotic systems. These kits cover several different types of chemistry, including salt extraction, filter plate-based solid-phase extraction and

Table 3.1 Overview of different automated DNA extraction systems

Instrument	Chemistry	Principle	Sample volume	Sample throughput	Remarks
Bioneer HT-Prep™ Automatic DNA Extractor	AccuPrep™ extraction kit	Solid-phase extraction (DNA bound to glass fibres, column washing steps)	0.5, 0.9 ml 1 × 96 10 × 96/day	96/1h	
Applied Biosystems ABI PRISM™ 6100 Nucleic Acid PrepStation[a]	Integrated semi-automated system	Solid-phase extraction	150 µl	96/~2h requires user intervention about every 15 min	Manual addition of reagents; quite lab-intensive; problem of clogged wells; extraction size very low by low DNA yield[a]
Corbett Life Science X-Tractor Gene RNA/DNA Extraction System	Kit adopters for Sigma, Qiagen, Promega, Macherey Nagel, Invitek	Depends on chemistry (bench top unit + vacuum station)		96/1h	
Gentra Systems Autopure LS®	PureGene® kits	Liquid-phase genomic DNA purification over protein + DNA precipitation	1–10 ml	96/8h	Complete sample tracking with bar codes + complete chain of custody

Instrument	Type	Chemistry	Volume	Time	Comments
Chemagen Module I[a]	Integrated system	Paramagnetic bead chemistry	1–200 µl 1–10 ml	96/40 min + 45 min manual preparation/run	To prevent cross-contamination the rods are shielded in plastic tips; performed very well; bar coding with integrated robot; instrument can cope with small + large sample volumes[a]
Roche MagNa Pure Compact[a] (larger instrument MagNA pure LC)	Integrated system	Paramagnetic bead chemistry	300 µl 1 ml	8/~25 min	Conveniently packaged foil-sealed cartridges; built in bar-code system for samples + reagents; UV lamp inside for decontamination; good DNA results[a]
Qiagen EZ1[a,b,c]	Integrated system EZ1 DNA kits	Paramagnetic bead chemistry	Cartridges for 200 and 350 µl fixed volume	6/~15 min	Predispensed, ready-to-use extraction reagents in conveniently packaged foil-sealed cartridges; no bar-coding facility; upgrade in near future; good DNA results[a,b]
Qiagen M48[c,d]	Integrated system MagAttract[®] DNA kits	Paramagnetic beads	1–500 µl	6/15 min 48/2 h	Good DNA results; UV lamp inside for decontamination; no detectable cross-contamination; no bar-coding facility; upgrade in near future[a,b]
Promega Maxwell[TM] 16 Instrument[c,e]	DNA IQ[TM]	Paramagnetic particles	Cartridges for high + low DNA content samples	16/20 min	Predispensed, ready-to-use extraction reagents in disposable cartridges, swabs or card punches were preprocessed; no detectable cross-contamination

Table 3.1 Continued

Instrument	Chemistry	Principle	Sample volume	Sample throughput	Remarks
Tecan Freedom EVO® 100[c]	DNA IQ™	Paramagnetic particles	96 swabs in 2.2 ml well plates	8 tip script/1.5 h	User can choose between aqueous + swab sample; blood samples were preprocessed, swabs + card punches were centrifuged; no detectable cross-contamination
	DRI charge switch technology[a]	Paramagnetic bead technology based on differential binding dependent on pH	10–20 µl 50–100 µl 1 ml	96/~2 h	Manual addition of reagents; bar-code possible; contamination in resuspended DNA; technical problems, require significant further development; poor PCR success rate; presence of protein[a]
	Macherey Nagel nucleospin[a]	Solid-phase extraction	200 µl up to 10 ml	96/~1 h	Manual addition of reagents; bar-code possible; problems with vacuum, require further development; one of best results; satisfactory DNA yield + purity[a]

					Comments
Promega MagneSil[a]		Paramagnetic beads	1–5 ml	8/~2 h	Manual addition of reagents; bar-code possible; problems not chemistry specific; needs extensive redesign; DNA yield quite poor, but good quality[a]
Autogen NA-3000EU[a]	Integrated system	Salt extraction	0.5–7 ml	48/~8 h	Simple to operate; addition of reagents + sample tracking must be performed manually, therefore risk of cross-contamination; no integrated bar-code system; 8% extraction failure rate[a]
Biomek® 2000 (3000 with new heater, shaker + deep-well plates[c,g])	DNA IQ[TM]	Paramagnetic particles			User can choose between aqueous + swab sample; blood samples were preprocessed, swabs + card punches were centrifuged; no detectable cross-contamination

[a] Evaluation report of automated extraction methodologies of the National Genetics Reference Laboratory (NGRL, Wessex, UK: http://www.ngrl.co.uk/Wessex/extraction.htm).
[b] Anslinger et al. (2005) and Montpetit et al. (2005).
[c] Forensic applications are known.
[d] Nagy et al. (2005) and Steinlechner et al. (2005).
[e] Bjerke (2006).
[f] Cowan (2006).
[g] Greenspoon et al. (2004), Crouse et al. (2005) and McLaren et al. (2006).

extractions based on paramagnetic beads. The systems were tested for a number of output factors, including DNA yield, purity and integrity, and suitability of DNA for downstream applications such as PCR. Anonymized EDTA blood samples and mock samples (water blanks) were used in the testing, and the survey also provided specifications for each system, including throughput, sample volume capacity and costs. Table 3.1 shows a summary of the results. Most of the systems did not have a bar-code system included in their design, so samples had to be tracked manually by means of tube/sample location. A final analysis of the survey results concluded that the integrated systems, which combined dedicated instruments and extraction chemistry, gave significantly better results than those that adapted existing DNA extraction kits to an automated platform using standard liquid-handling robotics. The study further concluded that although these kit-based chemistries could be optimized to a level satisfactory for diagnostic use, such systems would most likely require an extended optimization time to ensure good results. In terms of the different extraction chemistries tested, the solid-phase extractions gave the best results in spite of issues involving blocked wells and insufficient vacuum power, problems that are particular to this methodology. Overall, the simplest and most versatile systems appeared to be those based on paramagnetic bead chemistries.

Adapting a robotic extraction system for forensic casework can be a formidable problem, since the system has to be flexible enough to function efficiently across the enormous range of variation in DNA content obtained from a wide variety of forensic samples, such as blood, saliva, hair, vaginal swabs and contact stains on various carrier materials. The number of cells present in a sample can vary broadly – from more than 1000 nucleated cells in half a microlitre of blood to only a few cells in a sample obtained from a touch or a sneeze. Contact traces are becoming more and more prevalent in criminal cases, so the efficiency of extraction systems has to be rigorously successful down to picogram amounts of DNA while remaining exquisitely sensitive to the need to avoid potential contamination events.

Although there were early applications of automated extraction systems for clinical diagnostics, automated extraction systems for forensic evidentiary samples have only recently been reported (Greenspoon et al., 2004; Crouse et al., 2005) with the application of the BioMek® 2000 Workstation (Beckman Coulter, Inc., Fullerton, USA) combined with the DNA IQ™kit (Promega Corporation, Madison, WI, USA) for forensic casework samples. Nagy et al. (2005) have validated the Qiagen BioRobot M48 workstation (QIAGEN GmbH, Hilden, Germany) for nearly all types of forensic samples, and the smaller counterpart of this instrument, the Qiagen EZ1 (QIAGEN GmbH, Hilden, Germany), was shown to be valid for forensics work by Anslinger et al. (2005) and Montpetit et al. (2005). The DNA IQ™ system has been automated for the initial testing of forensic samples on robotic platforms such as the Tecan Freedom EVO® 100 (Cowan, 2006) as well as on the Maxwell™ 16 instrument (Bjerke et al., 2006). The Maxwell™ 16 instrument (Promega Corporation, Madison, US; Figure 3.3) operates differently than many other automated DNA purifica-

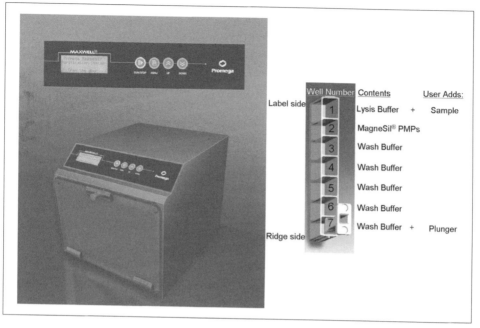

Figure 3.3 The Maxwell™ 16 instrument (Promega Corporation), with a close-up of the navigation liquid-crystal display and contents of prefilled reagent cartridges. Reproduced with permission from Promega Corporation

tion systems: rather than moving liquids from one well to another to carry out the various stages in DNA isolation and purification, the paramagnetic particles (PMPs) are moved from well to well during the purification process by individual magnets and disposable plungers.

Not surprisingly, all these described methods are based on the paramagnetic beads chemistry, which we and others have concluded is the simplest and most versatile system. However, the question of which automation system is best suited to forensic analysis remains open, and we now discuss three systems that are currently the most extensively validated robotic systems for forensics analysis: the Beckman BioMek® 2000, the Qiagen M48 and the Qiagen EZ1. All three systems use an extraction method that comprises the following steps (Figure 3.1c): cell lysis, binding of the DNA to the silica surface of paramagnetic particles in the presence of chaotropic reagents, washing steps to remove impurities and elution of the DNA.

The BioMek® 2000/DNA IQ™ system

This system (Figure 3.4) uses the DNA IQ™ kit-based chemistry: cells are briefly lysed either in a proteinase K-containing buffer or in the DNA IQ™ lysis buffer at either 57°, 68° or 95°C, depending on the substrate composition. For semen

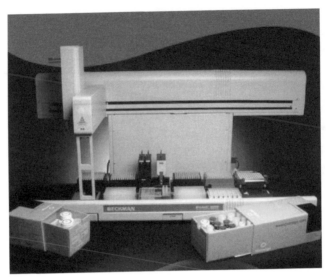

Figure 3.4 The DNA IQ™ system (Promega Corporation) with the BioMek® 2000 robot (Beckman Coulter). Reproduced with permission from Promega Corporation

stains differential extractions were performed. Lysates were then centrifuged prior to loading onto the robot. After initial contamination tests showed that a low level of contamination was introduced during the robotic extraction, the software method was modified to include automated resin addition, as well as the use of a 96-well deep-well plate and the replacement of the initial shaking step with a pipetting step (Greenspoon *et al.*, 2004). These changes are now included in the BioMek® 3000 Workstation (McLaren *et al.*, 2006). Greenspoon *et al.* (2004) have validated the BioMek® 2000/DNA IQ™ system by extensive contamination, efficiency and sensitivity tests described below. To detect contaminations a 'checkerboard' test was used: samples containing a concentrated source of DNA were loaded into wells with DNA alternating with reagent blanks across the whole plate, forming a checkerboard pattern.

To assess a system's efficiency at extracting low levels of DNA, a sensitivity study utilizing different dilutions of triplicated bloodstain punches (described below) was carried out: one sample was extracted by the automated procedure, one by the manual DNA IQ™ process and one by phenol–chloroform extraction. At dilutions of 1:10 and 1:100, all three methods produced similar yields of DNA. However, at a dilution of 1:1000 there was clearly a better yield from the BioMek® 2000/DNA IQ™ method than from manual extraction or from automated extraction using the PowerPlex® 1.1 or 16 BIO systems (Promega Corporation, Madison, USA). For sensitivity studies, blood from two different donors was deposited on various substrates (e.g. hand soap and lotion, carpet, black underwear, blue jeans, contraceptive foam, dirt, canvas). The aim of these different depositions was to study if inhibitory substances from the sample carrier might persist throughout DNA extraction and interfere with binding of DNA to the silica-coated paramagnetic resin. In this study, the STR profile was obtained for all of these depositions except for one donor's sample, which had been deposited on synthetic canvas. It seems likely that this failure resulted from the fact that the pre-heating step at 95°C can melt synthetic material and thus possibly destroyed the sample. However, STR profiles for synthetic canvas deposition were successfully obtained in the case of the second donor, indicating that these deposition conditions did not impose an absolute block to sample extraction; the exact reason for the failure in the first case is not known. DNA was successfully extracted from sexual assault samples, cigarette butts, blood stains, buccal swabs and various tissue samples, with no evidence of contamination throughout the extensive validation studies reported by Greenspoon *et al.* (2004). In addition, DNA extractions from manual pre-treated bone, hair and epithelial cells from touch evidence have been validated by Crouse *et al.* (2005).

The Qiagen BioRobot EZ1

The Qiagen EZ1 (Figure 3.5) is a compact benchtop instrument with its own integrated EZ1 DNA extraction chemistry designed to handle up to six samples in 15 minutes. The paramagnetic bead-based chemistry is conveniently packaged in a foil-sealed reagent cartridge (a robot-specific package), so the risk of contamination is extremely low. The instrument is controlled by a simple keypad and LCD screen, and different protocols are programmed by being loaded onto the instrument via cards that are plugged into a slot in the front of the instrument. The EZ1 was evaluated for DNA extraction from a variety of different evidence sample types, including blood, saliva and semen (Montpetit *et al.*, 2005) as well as cigarette butts, katagen and telogen hair roots, paraffin-embedded tissues and pulverized tooth (Anslinger *et al.*, 2005). Forensic samples were pre-treated in a variety of ways – some were treated with lysis buffer from the extraction kit and proteinase K and were incubated at 56°C for various times, while vaginal swabs with sperm cells were extracted by a differential lysis protocol (Figure 3.2b). Tooth and hair roots were lysed with a specific 'bone lysis buffer'. For all extractions, the EZ1 DNA Tissue kit was used in combination with the 'Forensic' programming card with the exception of embedded tissues, which used the 'Tissue' card instead. DNA yields were comparable to those from phenol–chloroform extraction, and the EZ1 purification process effectively removed PCR inhibitors. Variation in purification efficiency of the EZ1 DNA tissue kit reported by Anslinger *et al.* (2005) is no longer an issue. QIAGEN implemented functional Quality Control Testing procedures in order to assure good lot–lot consistency of purification efficiency for the EZ1 DNA Tissue Mini kits.

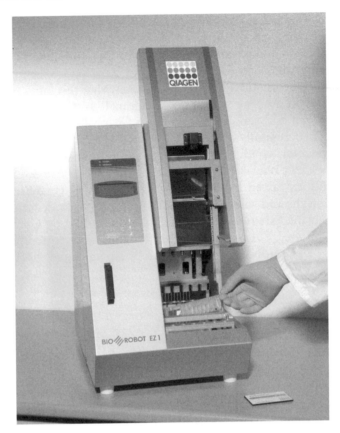

Figure 3.5 The Qiagen BioRobot EZ1. The EZ1 working deck has six slots to take the kit cartridges, an aligned tube rack to locate tubes for samples and re-suspended DNA and, above the deck, a robotic head with six positions, each having a spike for piercing the foil seals on the cartridges, a pipette and a magnetic probe. Reproduced with permission from QIAGEN Products

The Quiagen BioRobot M48

The Qiagen BioRobot M48 (Figure 3.6) is a dedicated DNA extraction instrument that was designed with its own integrated MagAttract® DNA extraction chemistry. This system can handle six samples in 15 minutes, and up to 48 samples in 2 hours. The workstation is equipped with special anti-contamination features, such as a completely enclosed robotic deck, a drop-catcher (as well as a control check for closing the door and cleaning the catcher), filter tips, a flat stainless-steel area and a UV sterilization system for decontamination between runs. After loading the instrument, there are no manual steps up until the point of collecting the pure DNA sample (Figures 3.6 and 3.7). The instrument is equipped with an interface that provides step-by-step instructions for

Figure 3.6 The Qiagen BioRobot M48. The workstation is equipped with special anti-contamination features, such as a completely enclosed robotic deck. Reproduced with permission from QIAGEN Products

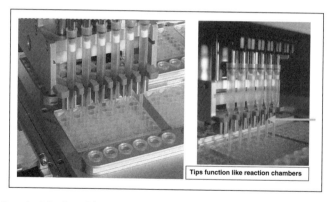

Tips function like reaction chambers

Figure 3.7 View inside the Qiagen BioRobot M48 workstation. Filter tips function like reaction chambers and lead to efficient purification. Reproduced with permission from QIAGEN Products

the run set-up as well as options for sample data import and export. Nagy *et al.* (2005) have worked out a simple, manageable manual protocol for the pre-treatment of forensic samples combined with an extraction protocol (using the Genomic DNA-Blood and Cells-Standard protocol) that is suitable for the full range of forensic specimen (blood samples, buccal swabs, blood stains, vaginal swabs, hairs, cigarette butts, bone meal, etc.). In their recommended standard lysis protocol, the forensic specimen was completely covered with detergent lysis buffer (50 mM Tris·HCl pH 7.4, 100 mM NaCl, 100 mM Na_2EDTA pH 8.1, 1% SDS) and incubated for 15 minutes at 90°C with thorough mixing. The only specimens for which modifications were necessary were telogen hair, sperm and bone meal. For telogen hair (i.e. hair that is in the dormant part of its growth cycle), TN_{CA} buffer (10 mM Tris·HCl pH 8.0, 100 mM NaCl, 1 mM $CaCl_2$, 2% SDS, 39 mM DTT, 250 µg/ml proteinase K; Hellmann *et al.*, 2001) was used instead of detergent lysis buffer. For sperm, differential extraction was not used (Figure 3.2c): 20 µl of 0.1 M DTT was generally added to 500 µl of detergent lysis buffer in the presence of sperm cells. For bone meal, lysis was preceded by multiple decalcification steps with 0.5 M EDTA followed by a last precise centrifugation step. Using these sample-specific protocols, complete DNA profiles were even obtained from most different types of bones ranging from 2 to 16 years old (macerated and high-density bones included), as well as from 9-year-old teeth. The only sample for which a profile was not obtained was the macerated skeleton part, for which the analysis resulted only in a positive amplification of the amelogenin system. Results showed that PCR inhibitors were efficiently removed from all of the forensic samples, and the prepared DNA remained stable after 2 years of storage. The BioRobot M48 preparative technology system was evaluated by Nagy *et al.* (2005) on the basis of a defined cell number model so that the DNA recovery rates could be exactly determined. In a 'real-world' analysis using unknown samples, the M-48 BioRobot® workstation has now been used in our laboratory for the extraction of DNA from more than 40 000 routine laboratory samples with a daily monitoring of the extraction efficiency. In addition, there has been no evidence of cross-contamination so far, as examined by PCR testing using 28–30 cycle reactions.

QIAGEN now offers automated DNA extraction kits specifically designed for both of their robotic systems (the EZ1 DNA Tissue Mini kit and the M48 MagAttract® DNA Mini kit). The company also offers protocols that enable processing of solid material directly in the sample tube with no need for centrifugation steps, so that high-quality DNA may be purified from a variety of forensic casework samples (*QIAGEN News*, 2006, e2). Specific pre-treatment protocols for the DNA extraction from various forensic samples are included.

In conclusion, the precision, sensitivity and reliability of the extraction systems discussed here have been extensively validated for application to forensic casework, and results demonstrate that these systems can produce DNA of high quality such that potential PCR inhibitors are removed and there is no need for any further purification steps. Overall, the automated DNA extraction systems discussed here have been shown to carry out safe and successful extractions for practically all types of forensic sample evidence, and should considerably simplify the reproducible processing of large numbers of samples.

3.6 References

Akane, A., Matsubara, K., Nakamura, H., Takahashi, S. and Kimura, K. (1994) Identification of the heme compound co-purified with deoxyribonucleic acid (DNA) from bloodstains. A major inhibitor of the polymerase chain reaction (PCR) amplification. *J. Forensic Sci.* **39**: 362–372.

Anslinger, K., Bayer, B., Rolf, B., Keil, W. and Eisenmenger, W. (2005) Application of the BioRobot EZ1 in a forensic laboratory. *Leg. Med.* 7: 164–168.

Bjerke, M., Kephart, D., Knox, C., Stencel, E., Krueger, S. and Lindsay, C. (2006) Forensic Applications of the Maxwell™ 16 Instrument. Promega Corporation. *Profiles DNA 9:* 3–5.

Butler, J.M. (2005) *Forensic DNA Typing. Biology, Technology, and Genetics of STR Markers* (2nd edn), Academic Press, New York, pp. 33–63.

Cowan, C. (2006) The DNA IQ™ System on the Tecan Freedom EVO® 100. Promega Corporation. *Profiles DNA 9:* 8–10.

Crouse, C., Yeung, S., Greenspoon, S., McGuckian, A., Sikorsky, J., Ban, J. and Mathies, R. (2005) Improving efficiency of a small forensic DNA laboratory: validation of robotic assays and evaluation of microcapillary array device. *Croat. Med. J.* 46: 563–577.

Gill, P., Jeffreys, A.J. and Werrett, D.J. (1985) Forensic application of DNA 'fingerprints'. *Nature* 318: 577–579.

Greenspoon, S.A., Ban, J.D., Sykes, K., Ballard, E.J., Edler, S.S., Baisden, M. and Covington, B.L. (2004) Application of the BioMek 2000 Laboratory Automation Workstation and the DNA IQ System to the extraction of forensic casework samples. *J. Forensic Sci.* 49: 29–39.

Hellmann, A., Rohleder, U., Schmitter, H. and Wittig, M. (2001) STR typing of human telogen hairs – a new approach. *Int. J. Legal Med.* 114: 269–273.

Holland, M.M., Cave, C.A., Holland, C.A. and Bille, T.W. (2003) Development of a quality, high throughput DNA analysis procedure for skeletal samples to assist with the identification of victims from the World Trade Center attacks. *Croat. Med. J.* 44: 259–263.

Holland, M.M., Fisher, D.L., Mitchell, L.G., Rodriquez, W.C., Canik, J.J., Merril, C.R. and Weedn, V.W. (1993) Mitochondrial DNA sequence analysis of human skeletal remains: identification of remains from the Vietnam War. *J. Forensic Sci.* 38: 542–553.

Höss, M. and Pääbo, S. (1993) DNA Extraction from Pleistocene bones by a silica-based purification method. *Nucleic Acids Res.* 21: 3913–3914.

Klintschar, M. and Neuhuber, F. (2000) Evaluation of an alkaline lysis method for the extraction of DNA from whole blood and forensic stains for STR analysis. *J. Forensic Sci.* 45: 669–673.

McHale, R., Stapleton, P.M. and Bergquist, P.L. (1991) A rapid method for the preparation of samples for PCR. *Biotechniques* 10: 20–22.

McLaren, B., Bjerke, M. and Tereba, A. (2006) Automating the DNA IQ™ System on the Biomek® 3000 Laboratory Automation Workstation. Promega Corporation. *Profiles DNA 9:* 11–13.

Miller, S.A., Dykes, D.D. and Polesky, H.F. (1988) A simple salting out procedure for extracting DNA from human nucleated cells. *Nucleic Acids Res.* 10: 1215.

Montpetit, S.A., Fitch, I.T. and O'Donnell, P.T. (2005) A simple automated instrument for DNA extraction in forensic casework. *J. Forensic Sci.* 50: 553–563.

Moreira, D. (1998) Efficient removal of PCR inhibitors using agarose-embedded DNA preparations. *Nucleic Acids Res.* 26: 3309–3310.

Moss, D., Harbison, S.A. and Saul, D.J. (2003) An easily automated, closed-tube forensic DNA extraction procedure using a thermostable proteinase. *Int. J. Legal Med.* 117: 340–349.

Nagy, M., Otremba, P., Krüger, C., Bergner-Greiner, S., Anders, P., Henske, B., Prinz, M. and Roewer, L. (2005) Optimization and validation of a fully automated silica-coated magnetic beads purification technology in forensics. *Forensic Sci. Int.* **152**: 13–22.

Parson, W., Niederstätter, H., Brandstätter, A. and Berger, B. (2003) Improved specifity of Y-STR typing in DNA mixture samples. *Int. J. Legal Med.* **117**: 109–114.

Prado, V.F., Castro, A.K., Oliveira, C.L., Souza, K.T. and Pena, S.D. (1997) Extraction of DNA from human skeletal remains: practical applications in forensic sciences. *Genet. Anal.* **14**: 41–44.

Prinz, M., Ishii, A., Coleman, A., Baum, H.J. and Shaler, R.C. (2001) Validation and casework application of a Y chromosome specific STR multiplex. *Forensic Sci. Int.* **120**: 177–188.

Sambrook, J., Fritsch, E.F. and Maniatis, T. (1989) In: *Molecular Cloning: A Laboratory Manual* (2nd edn), Cold Spring Harbor Laboratory Press, Plainview, NY, 1989, pp. E.3–E.4, E.10.

Starnbach, M.N., Falkow, S. and Tomkins, L.S. (1989) Species-specific detection of Legionella pneumophila in water by DNA amplification and hybridization. *J. Clin. Microbiol.* **27**: 1257–1261.

Steinlechner, M., Parson, W., Rabl, W., Grubwieser, P. and Scheithauer, R. (2005) DNS-Laborstrategie zur Identifizierung von Katastrophenopfern. *Rechtsmedizin* **15**: 473–478.

Walsh, P.S., Metzger, D. and Higuchi, R. (1991) Chelex 100 as a medium for simple extraction of DNA for PCR-based typing from forensic material. *Biotechniques* **10**: 506–513.

Wilson, I.G. (1997) Inhibition and facilitation of nucleic acid amplification. *Appl. Environ. Microbiol.* **63**: 3741–3751.

4

Real-time quantitative PCR in forensic science

Antonio Alonso and Oscar García

4.1 Introduction

The specific quantification of human DNA from forensic evidentiary materials is a recommended procedure for a reliable short tandem repeat (STR) analysis, which is the current gold standard for DNA profiling in forensic casework (Jobling and Gill, 2004) (see Chapter 5). An estimation of the DNA quantity is made to adjust the DNA input of subsequent end-point polymerase chain reaction (PCR)-based DNA profiling methods, preventing PCR failures that are due to the absence of DNA, and avoiding PCR artefacts such as random allele dropout produced by stochastic amplification effects from low copy number (LCN) DNA samples (under 100 pg of DNA) (Gill *et al.*, 2000), and also preventing off-scale over-amplification artefacts (including $n + 1$ peaks, increased stutter bands and pullout) associated with an excess of DNA input in the PCR. In addition, an accurate DNA quantification helps to prevent the unnecessary waste of DNA, which is especially important when analysing LCN DNA samples.

In the past, forensic laboratories have used the slot-blot hybridization approach to target the D17Z1 locus (Waye and Willard, 1986) – a highly repetitive alphoid primate-specific sequence – for DNA quantification in forensic casework (Waye *et al.*, 1989). However, this methodology, with a detection limit above the limit of the STR profiling approaches, was often not sensitive enough to detect LCN forensic DNA samples. In addition, the method is labour-intensive, time-consuming and poorly suited to high-throughput sample flow.

Several studies (Andréasson *et al.*, 2002; Alonso *et al.*, 2003; Nicklas and Buel, 2003; Richard *et al.*, 2003; Walker *et al.*, 2003, 2005; Alonso *et al.*, 2004;

Molecular Forensics. Edited by Ralph Rapley and David Whitehouse
Copyright 2007 by John Wiley & Sons, Ltd.

Timken *et al.*, 2005; Andréasson *et al.*, 2006; Swango *et al.*, 2006) have demonstrated the usefulness of real-time PCR for a sensitive, specific and high-throughput quantification of both human nuclear DNA (nuDNA) and mitochondrial DNA (mtDNA) in forensics and ancient DNA studies.

The recent development of a commercially available real-time PCR human DNA quantification kit (Green *et al.*, 2005) has also contributed to a worldwide use of real-time PCR in forensic genetics.

In this chapter we will review all the different real-time PCR assays that have been applied in forensic genetics for the specific quantification of human autosomal, X and Y chromosome and mitochondrial DNA targets. We also review some real-time PCR assays for rapid quantification of non-human species of interest in forensic and ancient DNA studies, including the identification of pathogens in microbial forensics (Budowle *et al.*, 2005). Other applications of real-time PCR assays, such as allele discrimination and single nucleotide polymorphism genotyping, are outside of the scope of this review.

4.2   Current real-time PCR chemistries

There are two general fluorogenic methods to monitor the real-time progress of the PCR: by measuring *Taq* polymerase activity using double-stranded DNA binding dye chemistry (SYBR Green or ethidium bromide) (Higuchi *et al.*, 1993) or by measuring the 5′-nuclease activity of the *Taq* DNA polymerase to cleave a target-specific fluorogenic probe (a TaqMan probe: an oligonucleotide complementary to a segment of the template DNA, with both a reporter and a quencher dye attached, that only emits its characteristic fluorescence after cleavage) (Holland *et al.*, 1991) Although TaqMan assays are the most popular probe-based assays in forensic laboratories, alternative probe-based chemistry, such as molecular beacons (Tyagi and Kramer, 1996) or scorpion primers and probes (Whitcombe *et al.*, 1999), could also be employed for specific target detection.

Real-time analysis of the fluorescence levels at each cycle of the PCR (amplification plot) allows a complete picture to be obtained of the whole amplification process for each sample. In the initial cycles of the PCR a baseline is observed without any significant change in fluorescence signal. An increase in fluorescence above the baseline indicates the detection of accumulated PCR product. The higher the initial input of the target genomic DNA, the sooner a significant increase in fluorescence is observed. The cycle at which fluorescence reaches an arbitrary threshold level during the exponential phase of the PCR is named *Ct* (threshold cycle). A standard curve can be generated by plotting the log of the starting DNA template amount of a set of previously quantified DNA standards against their *Ct* values. Therefore, an accurate estimation of the starting DNA amount from unknown samples is accomplished by comparison of the measured *Ct* values with the *Ct* values of the standard curve.

Compared to end-point PCR quantification methods, the use of *Ct* values is a more reliable quantification assay. This is mainly due to the fact that *Ct* determination is performed during the high-precision exponential phase of the PCR when none of the reaction components are limiting, contrary to PCR end-point measurements.

Although SYBR Green assay provides the simplest and most economical format for detecting and quantifying PCR products in real-time reactions, the main limitation is that non-specific amplifications (primer-dimer, non-human products, etc.) cannot be distinguished from specific amplifications. On the other hand, the amplicon/dye ratio varies with amplicon length. In addition, SYBR Green can only be used in single PCR reactions, therefore the use of this assay should be restricted to optimized PCR reactions producing single PCR products free from non-specific PCR artefacts.

The probe-based real-time PCR assay has been the method of choice to quantify human nuDNA and mtDNA in forensic genetics because of its superior specificity and quantitation accuracy in comparison with SYBR Green assays (see Chapter 9). Another advantage of probe-based methods is the feasibility to perform multiplex PCR of different targets (Andréasson *et al.*, 2002; Timken *et al.*, 2005; Walker *et al.*, 2005). Probe-based assays also provide the possibility to perform, in a single PCR reaction, not only specific human DNA quantification but also different qualitative analyses, such as gender determination (Alonso *et al.*, 2003, 2004; Andréasson and Allen, 2003; Green *et al.*, 2005), DNA degradation (Alonso *et al.*, 2004; Swango *et al.*, 2006) and *Taq* inhibition rate (Timken *et al.*, 2005; Green *et al.*, 2006; Swango *et al.*, 2006).

4.3   Human nuclear DNA quantification

Single-copy autosomal targets

The use of probe-based real-time PCR to quantify human nuclear DNA in forensic analysis was firstly described by Andréasson *et al.* (2002). A 78 bp region of the human retinoblastoma susceptibility gene (RB1), a nuclear-encoded single-copy gene located on chromosome 13, was the target in a multiplex PCR quantification assay that was also designed to amplify an mtDNA target (see autosomal DNA and mtDNA quantification by a duplex PCR assay in Table 4.1). The system has been shown to detect down to nuDNA single copies in the dilution series of the standard curve and has been applied to quantify nuDNA from different forensic specimens such as skin debris, saliva stains, hair and bloodstains. This assay has been used recently to quantify nuDNA in the roots and distal sections of plucked and shed head hairs and also from fingerprints and accessories (Andréasson *et al.*, 2006). However, the choice of RB1 as a real-time quantitative PCR target may not be ideal because the RB1 sequence

Table 4.1 Real-time PCR assays used in forensic genetics for specific quantification of human DNA targets

Assay type	Marker (size)	Chemistry[a]	Comments	References
Single-copy autosomal DNA quantification by singleplex PCR	HUMTHO1 (62bp)	FAM–TaqMan-MGB probe	Targeting of a conserve region (out of the STR array) of the HumTHO1 locus	(Richard et al., 2003)
Single-copy autosomal DNA quantification by duplex PCR	hTERT (62bp)	FAM–TaqMan-MGB probe (VIC–TaqMan-MGB probe for IPC detection)	Includes an internal standard to evaluate Taq inhibition Commercially available kit	(Green et al., 2005)
	TH01 (170–190bp) CSF1PO (67bp)	FAM–TaqMan-MGB probe VIC–TaqMan-MGB probe (NED–TaqMan-MGB probe for IPC detection)	The ratio of nuCSF:nuTHO1 quantifications was shown to provide a good estimation of the degree of degradation Includes an internal standard to evaluate Taq inhibition	(Swango et al., 2006)
Single-copy X and Y chromosome DNA quantification by singleplex PCR	AMEL X (106bp) AMEL Y (112bp)	FAM–TaqMan-MGB probe VIC–TaqMan-MGB probe	Simultaneous nDNA quantification and gender determination	(Alonso and Matin, 2005; Alonso et al., 2003, 2004)
	AMEL X/Y (70–73bp)	SYBR-Green	nDNA quantification and gender determination with certain limitations	(Andréasson and Allen, 2003)
	SRY (61–64bp)	FAM–TaqMan-MGB probe (VIC–TaqMan-MGB probe for IPC detection)	Specific quantification of Y chromosome copy number. Includes an internal standard to evaluate Taq inhibition	(Green et al., 2005)

Method	Target	Detection	Description	Reference
Quantification of Alu sequence targets by singleplex PCR	Alu Sequence (124bp)	SYBR Green		(Nicklas and Buel, 2003) (Walker et al., 2003)
	Inter-Alu sequences (pool of different sizes)	SYBR Green		
	Intra-Alu sequences (200–226bp)	SYBR Green		
MtDNA quantification by singleplex PCR	HV1 (113bp)	VIC–Taqman-MGB probe	DNA degradation assessment by simultaneous quantification of two different sized fragments	(Alonso et al., 2004)
	HV1 (287bp)	FAM–Taqman-MGB probe		
Autosomal DNA and mtDNA quantification by duplex PCR	RB1 (78bp)	FAM–TaqMan probe	Duplex PCR amplification for simultaneous quantification of nuDNA and mtDNA	(Andréasson et al., 2002, 2006) (Timken et al., 2005)
	mtDNA coding region (142bp)	VIC–TaqMan probe		
	HUMTH01 (170–190bp)	FAM–TaqMan probe	Duplex PCR amplification for simultaneous quantification of nuDNA and mtDNA	
	MtDNA region ND1 (69bp)	VIC–Taqman-MGB probe (NED–TaqMan probe for IPC detection)	Includes an internal standard to evaluate Taq inhibition	
Autosomal DNA, Y chromosome and mtDNA quantification by triplex PCR	Nuclear Alu sequence (71bp)	VIC–Taqman-MGB probe	Triplex PCR amplification for simultaneous quantification of nuDNA, mtDNA and male Y-DNA	(Walker et al. 2005)
	Y-Chr target (69bp)	NED–Taqman-MGB probe		
	mtDNA target (79bp)	FAM–Taqman-MGB probe		

[a] FAM, 6-carboxyfluorescein; NED, 2'-chloro-5'-fluoro-7',8'-fused phenyl-1,4-dichloro-6-carboxyfluorescein; VIC, 2'-chloro-7'-phenyl-1,4-dichloro-6-carboxyfluorescein.

is relatively conserved among different species and perhaps it is not sufficiently primate-specific to be of general forensic utility (Timken *et al.*, 2005).

A TaqMan-MGB (minus groove binder) real-time PCR design has been developed to target a small region of 62 bp located 31 bp downstream from the polymorphic repeat region of the HumTH01 locus. The assay has been applied to the quantification of human nuDNA from a variety of body fluid stains (Richard *et al.*, 2003).

A human DNA quantification kit was developed for the quantification of human nuDNA by targeting a 62 bp portion of the human telomerase reverse transcriptase (hTERT) locus using a TaqMan-MGB assay that includes an internal PCR control (IPC) for the assessment of PCR efficiency against *Taq* inhibitors (Green *et al.*, 2005). The kit has been validated for use in forensic casework according to the Scientific Working Group on DNA Analysis Methods (SWGDAM) guidelines and now is the most used assay in forensic casework for nuDNA quantitation.

More recently, a multiplex quantitative PCR assay has been described to amplify simultaneously two target sequences of different length, the TH01 STR locus (170–190 bp) and the upstream flanking region of the CSF1PO STR locus (67 bp), which allows for the assessment of DNA degradation in samples of forensic interest (Swango *et al.*, 2006). The assay also includes an internal PCR control target sequence to allow for an assessment of PCR inhibition.

Alu Repetitive elements

Several real-time PCR assays have also been developed to target different *Alu* repetitive sequences for highly sensitive nuDNA quantification. Nicklas and Buel (2003) have described a SYBR Green real-time PCR assay using specific primers to target a 124 bp *Alu* sequence with primate specificity and 1 pg sensitivity. Walker *et al.* (2003) presented two alternative SYBR Green real-time PCR designs to target inter-*Alu* (including a complex pool of sequences of different sizes) and intra-*Alu* (200–226 bp) sequences, respectively. Inter-*Alu* assay was found to be not completely human-specific while intra-*Alu* PCR was demonstrated to be a specific and sensitive method for human nuDNA quantification in the range of 10 ng to 1 pg.

The main advantages of SYBR Green real-time PCR designs to target *Alu* assays are simplicity and high sensitivity. The disadvantages are the necessity of optimization to avoid the possibility of unspecific PCR artefacts and the possible inaccuracy in the DNA quantification as a consequence of a hypothetical individual variation in the copy number of *Alu* sequences.

Walker *et al.* (2005) have described a TaqMan-MGB triplex real-time PCR assay for simultaneous quantitation of human nuDNA (based on an intra-*Alu* sequence design), mtDNA and Y chromosome DNA (see Table 4.1, Autosomal DNA, Y chromosome and mtDNA quantification by triplex PCR).

X and Y chromosome targets

Alonso *et al.* (2003, 2004; Alonso and Martin, 2005) described a method for nuDNA quantification based on TaqMan-MGB real-time PCR amplification of a segment of the X–Y homologous amelogenin (AMG) gene that allowed the simultaneous estimation of a Y-specific fragment (AMGY: 112 bp) and an X-specific fragment (AMGX: 106 bp), making possible not only DNA quantitation but also sex determination. Detection of the specific AMGX fragment (106 bp) and AMGY fragment (112 bp) was achieved using the primer pair sequences previously described (Sullivan *et al.*, 1993) and two fluorogenic MGB probes that specifically detect the AMGX fragment (FAM-labelled) or the AMGY fragment (VIC-labelled). The MGB probes were designed to target the 6 bp X-deletion / Y-insertion segment within the AMG second intron fragment. The method has been applied to the analysis of LCN DNA samples in forensic and ancient DNA studies (Alonso *et al.*, 2004).

A SYBR-green real-time PCR assay of the human amelogenin gene using specific primers to produce a Y-specific fragment of 73 bp and a 3-bp-deleted X-specific fragment of 70 bp was described (Andréasson and Allen, 2003). The assay allows quantification of the nuDNA copy number, but sex determination, which is based on a dissociation curve analysis that displays the different melting temperatures of X- and Y-specific products, has certain limitations for forensic applications and especially for the analysis of mixed male–female forensic samples.

A specific real-time PCR quantification kit based on the TaqMan-MGB detection of a region (61–64 bp) of the SRY locus is also commercially available (Green *et al.*, 2005). The assay, which detects only male DNA, is intended particularly for use in samples with mixed male–female DNA, such as sexual assault evidence, where it may be useful for specific male DNA detection. The kit has also been validated for use in forensic casework according to SWGDAM guidelines.

A sex chromosome TaqMan-MGB assay was designed around a 90 bp deletion on the X chromosome in an X–Y homologous region to target a 77 bp fragment on the human X chromosome and a 167 bp fragment on the human Y chromosome (Walker *et al.*, 2005).

4.4 Human mitochondrial DNA quantification

The use of a TaqMan real-time PCR assay for quantification of mitochondrial human DNA from forensic specimens was first described by Andréasson *et al.* (2003) by targeting a 142 bp region spanning over the genes for tRNA lysine and ATP Synthase 8 that can be amplified in a single PCR reaction or in combination with an nuDNA target. The assay was tested on 236 forensic specimens (hair, bloodstains, fingerprints, skin debris, saliva stains and others) containing from zero to >100 000 mtDNA copies and more recently on different LCN DNA samples such as fingerprints (Andréasson *et al.*, 2006).

Alonso *et al.* (2003, 2004) reported the specific quantification of human mtDNA by monitoring the real-time progress of the PCR amplification of two different fragment sizes (113 bp and 287 bp) within the hypervariable region 1 (HV1) of the mtDNA control region, using two fluorogenic probes to specifically determine the mtDNA copy of each fragment size category. This additional information – number of copies in each size category – has been demonstrated to be very helpful to evaluate the mtDNA preservation state from ancient bone samples.

A 69 bp fragment of the mtDNA NADH dehydrogenase subunit 1(ND1) locus has also been used as a target for mtDNA quantitation in forensic specimens using a TaqMan duplex real-time PCR assay that allows simultaneous quantification of human nuDNA (Timken *et al.*, 2005). Another mtDNA target used in forensics by TaqMan-MGB assays is a 79 bp fragment of a conserved region of the human mtDNA, which is co-amplified with an autosomal and a Y chromosome target (Walker *et al.*, 2005).

Recently a novel forensic use of quantitative real-time PCR (rtPCR) using TaqMan-MGB probes has been described, targeting the highly variable mitochondrial single nucleotide polymorphism 16519T/C to investigate heteroplasmic mixtures with an accurate quantification of the minor allele down to 9% (Niederstatter *et al.*, 2006).

4.5   Detection and quantification of non-human species

Forensic analysis of non-human evidence is gaining importance and becoming widely used in forensic laboratories for the identification of animal material recovered from the crime scene (usually pet hairs). A SYBR Green quantitative real-time PCR assay has been developed for the quantification of genomic DNA extracted from domestic cat samples (Menotti-Raymond *et al.*, 2003)

Investigation of the illegal trade in endangered species is another field of application of real-time quantitative PCR. A highly sensitive tiger-specific real-time PCR assay has been described using primers specific to the tiger mitochondrial cytochrome *b* gene. Successful amplification has been demonstrated from blood, hair and bone as well as from a range of traditional Chinese medicines spiked with 0.5% tiger bone (Wetton *et al.*, 2004).

Real-time quantitative PCR is also an emerging technique in ancient DNA studies of non-human species. Poinar *et al.* (2003) quantitated the number of mitochondrial 16S rDNA copies for fragments of three different lengths (114/252/522 bp) by using a TaqMan real-time PCR assay from ancient coprolite remains, demonstrating that there was roughly a 100-fold drop in the number of amplifiable mtDNA molecules for every doubling in amplification length.

Quantitative PCR analysis of DNA from non-invasive samples with low DNA content is also a useful tool for molecular characterization of wild animal populations in molecular ecology, including nuDNA quantification (Morin *et al.*, 2001) and sex determination (Morin *et al.*, 2005).

Real-time quantitative PCR is also a rapid and highly sensitive methodological tool for detection, identification and individualization of microbial agents that could be used in bioterrorist acts. For instance, real-time PCR assays are routinely used to detect the presence of DNA from *Bacillus anthracis* in environmental samples by using both unique plasmid-borne and chromosomal genes (Jones *et al.*, 2005). Other highly sensitive real-time PCR assays for the detection of potential bioterrorism agents such as *Yersinia Pestis*, *Francisella tularensis*, *Brucella* spp. and *Burkholderia* spp. have also been developed (see a review of the assays for biodefence used at the USA Armed Forces Institute of Pathology in Jones *et al.*, 2005).

4.6 Concluding remarks and perspectives

Real-time quantitative PCR has become a widely used technique for sensitive and specific quantification of both human nuDNA and mtDNA in forensics and ancient DNA studies, offering several advantages with respect to other current methodologies (hybridization or end-point PCR methods), including: higher sensitivity and dynamic range of quantitation, unnecessary post-PCR processing, automation feasibility and high throughput, and the possibility to simultaneously perform different qualitative analyses (gender determination, mtDNA degradation, *Taq* inhibition rate, etc.).

However, the data obtained by comparing different real-time PCR methods across different laboratories (Kline *et al.*, 2005) show differences with regard to precision and bias. These results emphasize the need to develop a standard reference material for DNA quantification in the forensic field (Kline *et al.*, 2005).

Real-time PCR methods will continue to take advantage of technical developments to improve the multiplexing capability, the automation feasibility and the detection limit.

The use of real-time quantitative PCR methods for body fluid (saliva, semen and blood) identification by targeting messenger RNA markers (Nussbaumer *et al.*, 2006) is a novel molecular approach within the forensic field that will probably increase in use and importance. The forensic application of real-time quantitative PCR assays to target other markers of gene expression remains to be explored.

4.7 References

Alonso, A. and Martin, P. (2005) A real-time PCR protocol to determine the number of amelogenin (X–Y) gene copies from forensic DNA samples. *Methods Mol. Biol.* **297**: 31–44.

Alonso, A., Martin, P., Albarran, C., Garcia, P., Primorac, D., Garcia, O., Fernandez de Simon, L., Garcia-Hirschfeld, J., Sancho, M. and Fernandez-Piqueras, J. (2003) Specific quantification of human genomes from low copy number DNA samples in forensic and ancient DNA studies. *Croat. Med. J.* **44**: 273–280.

Alonso, A., Martin, P., Albarran, C., Garcia, P., Garcia, O., de Simon, L.F., Garcia-Hirschfeld, J., Sancho, M., de La Rua, C. and Fernandez-Piqueras, J. (2004) Real-time PCR designs to estimate nuclear and mitochondrial DNA copy number in forensic and ancient DNA studies. *Forensic Sci. Int.* **139**: 141–149.

Andréasson, H. and Allen, M. (2003) Rapid quantification and sex determination of forensic evidence materials. *J. Forensic Sci.* **48**: 1280–1287.

Andréasson, H., Gyllensten, U. and Allen, M. (2002) Real-time DNA quantification of nuclear and mitochondrial DNA in forensic analysis. *BioTechniques* **33**: 402–411.

Andréasson, H., Nilsson, M., Budowle, B., Lundberg, H. and Allen, M. (2006) Nuclear and mitochondrial DNA quantification of various forensic materials. *Forensic Sci. Int.* **164**: 56–64.

Budowle, B., Johnson, M.D., Fraser, C.M., Leighton, T.J., Murch, R.S. and Chakraborty, R. (2005) Genetic analysis and attribution of microbial forensics evidence. *Crit. Rev. Microbiol.* **31**: 233–254.

Gill, P., Whitaker, J., Flaxman, C., Brown, N. and Buckelton, J. (2000) An investigation of the rigor of interpretation rules for STRs derived from less than 100 pg of DNA. *Forensic Sci. Int.* **112**: 17–40.

Green, R.L., Roinestad, I.C., Boland, C. and Hennessy, L.K. (2005) Developmental validation of the quantifiler real-time PCR kits for the quantification of human nuclear DNA samples. *J. Forensic Sci.* **50**: 809–825.

Higuchi, R., Fockler, C., Dollinger, G. and Watson, R. (1993) Kinetic PCR: Real-time monitoring of DNA amplification reactions. *Biotechnology* **11**: 1026–1030.

Holland, P.M., Abramson, R.D., Watson, R. and Gelfand, H. (1991) Detection of specific polymerase chain reaction product by utilizing the 5′-3′exonuclease activity of *Thermus Aquaticus* DNA polymerase. *Proc. Natl. Acad. Sci. USA* **88**: 7276–7280.

Jobling, M.A. and Gill, P. (2004) Encoded evidence: DNA in forensic analysis. *Nat. Rev. Genet.* **5**: 739–751.

Jones, S.W., Dobson, M.E., Francesconi, S.C., Schoske, R. and Crawford, R. (2005) DNA assays for detection, identification, and individualization of select agent microorganisms. *Croat. Med. J.* **46**: 522–529.

Kline, M.C., Duewer, D.L., Redman, J.W. and Butler, J.M. (2005) Results from the NIST 2004 DNA Quantitation Study. *J. Forensic Sci.* **50**: 570–578.

Menotti-Raymond, M., David, V., Wachter, L., Yuhki, N. and O'Brien, S.J. (2003) Quantitative polymerase chain reaction-based assay for estimating DNA yield extracted from domestic cat specimens. *Croat. Med. J.* **44**: 327–331.

Morin, P.A., Chambers, K.E., Boesch, C. and Vigilant, L. (2001) Quantitative PCR analysis of DNA from noninvasive samples for accurate microsatellite genotyping of wild chimpanzees (*Pan troglodytes*). *Mol. Ecol.* **10**: 1835–1844.

Morin, P.A., Nestler, A., Rubio-Cisneros, N.T., Robertson, K.M. and Mesnick, S.L. (2005) Interfamilial characterization of a region of the ZFX and ZFY genes facilitates sex determination in cetaceans and other mammals. *Mol. Ecol.* **14**: 3275–3286.

Niederstatter, H., Coble, M.D., Grubwieser, P., Parsons, T.J. and Parson, W. (2006) Characterization of mtDNA SNP typing and mixture ratio assessment with simultaneous real-time PCR quantification of both allelic states. *Int. J. Legal. Med.* **120**: 18–23.

Nicklas, J.A. and Buel, E. (2003) Development of an Alu-based, real-time PCR method for quantitation of human DNA in forensic samples. *J. Forensic Sci.* **48**: 936–944.

Nussbaumer, C., Gharehbaghi-Schnell, E. and Korschineck, I. (2006) Messenger RNA profiling: a novel method for body fluid identification by real-time PCR. *Forensic Sci. Int.* **157**: 181–186.

Poinar, H., Kuch, M., McDonald, G., Martin, P. and Pääbo, S. (2003) Nuclear gene sequences from a Late Pleistocene sloth coprolite. *Curr. Bio.* **13**: 1150–1152.

Richard, M.L., Frappier, R.H. and Newman, J.C. (2003) Developmental validation of a real-time quantitative PCR assay for automated quantification of human DNA. *J. Forensic Sci.* **48**: 1041–1046.

Sullivan, K.M., Mannucci, A., Kimpton, C.P. and Gill, P. (1993) A rapid and quantitative DNA sex test: fluorescence-based PCR analysis of X-Y homologous gene amelogenin. *BioTechniques* **15**: 637–641.

Swango, K.L., Timken, M.D., Chong, M.D. and Buoncristiani, M.R. (2006) A quantitative PCR assay for the assessment of DNA degradation in forensic samples. *Forensic Sci. Int.* **158**: 14–26.

Timken, M.D., Swango, K.L., Orrego, C. and Buoncristiani, M.R. (2005) A duplex real-time qPCR assay for the quantification of human nuclear and mitochondrial DNA in forensic samples: implications for quantifying DNA in degraded samples. *J. Forensic Sci.* **50**: 1044–1060.

Tyagi, S. and Kramer, F.R. (1996) Molecular beacons: probes that fluoresce upon hybridization. *Nat. Biotechnol.* **14**: 303–308.

Walker, J.A., Hedges, D.J., Perodeau, B.P., Landry, K.E., Stoilova, N., Laborde, M.E., Shewale, J., Sinha, S.K. and Batzer, M.A. (2005) Multiplex polymerase chain reaction for simultaneous quantitation of human nuclear, mitochondrial, and male Y-chromosome DNA: application in human identification. *Anal. Biochem.* **337**: 89–97.

Walker, J.A., Kilroy, G.E., Xing, J., Shewale, J., Sinha, S.K. and Batzer, M.A. (2003) Human DNA quantitation using Alu element-based polymerase chain reaction. *Anal. Biochem.* **315**: 122–128.

Waye, J.S., Presley, L.A., Budowle, B., Shutler, G.G. and Fourney, R.M. (1989) A simple and sensitive method for quantifying human genomic DNA in forensic specimen extracts. *Biotechniques* **7**: 852–855.

Waye, J.S. and Willard, H.F. (1986) Structure, organization, and sequence of alpha satellite DNA from human chromosome 17: evidence for evolution by unequal crossing-over and an ancestral pentamer repeat shared with the human X chromosome. *Mol. Cell Biol.* **6**: 3156–3165.

Wetton, J.H., Tsang, C.S., Roney, C.A. and Spriggs, A.C. (2004) An extremely sensitive species-specific ARMS PCR test for the presence of tiger bone DNA. *Forensic Sci. Int.* **140**: 139–145.

Whitcombe, D., Theaker, J., Guy, S.P., Brown, T. and Little, S. (1999) Detection of PCR products using self-probing amplicons and fluorescence. *Nat. Biotech.* **17**: 804–807.

5

Minisatellite and microsatellite DNA typing analysis

Keiji Tamaki

5.1 Introduction

Over 20 years have passed since the development of DNA fingerprinting, which marked the beginning of forensic DNA typing. Since then, human tandem repeat DNA sequences have become the preferred choice for forensic DNA analysis. These repeat sequences are classified into minisatellites (variable number tandem repeats, VNTRs) and microsatellites (short tandem repeats, STRs). In this chapter, we will discuss the historical and current forensic applications of such tandem repeats.

5.2 Minisatellites

'DNA fingerprinting' using multi-locus probes (MLPs)

Roughly 3% of the human genome is comprised of tandem repeats (International Human Genome Sequencing Consortium, 2001) that – with the exclusion of satellite DNA – can be classified into two groups according to the size of the repeat unit and overall length of the repeat array. Human minisatellites or VNTR loci have repeat units ranging in length from 6 bp to more than 100 bp depending on the locus, with arrays usually kilobases in length. Human GC-rich minisatellites are preferentially found clustered in the recombination-proficient subtelomeric regions of chromosomes (Royle *et al.*, 1988). Some minisatellite loci show very high levels of allele length variability.

Molecular Forensics.   Edited by Ralph Rapley and David Whitehouse
Copyright 2007 by John Wiley & Sons, Ltd.

The accidental discovery of hypervariable minisatellite loci detectable with MLPs in 1984 by Sir Alec Jeffreys launched a new age in forensic investigation (Jeffreys *et al.*, 1985a). These minisatellites were detected by hybridization of probes to Southern blots of restriction-enzyme-digested genomic DNA, to reveal restriction fragment length polymorphisms (RFLPs). A common 10–15 bp 'core' GC-rich sequence shared between different minisatellite loci allowed MLPs to detect a wide array of minisatellites simultaneously, producing multi-band (barcode-like) patterns known as 'DNA fingerprints' (Figure 5.1).

Using only a single MLP designated 33.15, the match probability between unrelated individuals was estimated at $<3 \times 10^{-11}$, and in the case of two MLPs (33.15 and 33.6) that detect different sets of minisatellites the match probability is $<5 \times 10^{-19}$ (Jeffreys *et al.*, 1985b). These probabilities are so low that the only individuals possessing identical DNA fingerprints are monozygotic twins. The

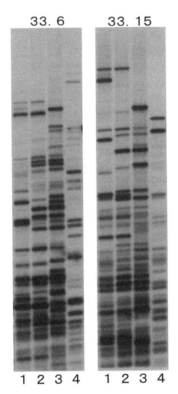

Figure 5.1   Examples of DNA fingerprint autoradiographs using multi-locus probes 33.6 and 33.15 obtained from a paternity test. The barcode-like patterns of mother (lane 1) and child (lane 2) are compared with two alleged fathers (lanes 3 and 4). One of the alleged fathers (lane 3) has not been excluded as the biological father because he possesses all the similar-sized bands inherent in the child that were not inherited from the mother. The remaining candidate (lane 4) was excluded from paternity due to a lack of common bands with the child

MLPs have proven their use in paternity testing (Jeffreys *et al.*, 1991a) and immigration cases (Jeffreys *et al.*, 1985c). However, a significant drawback associated with MLPs is the requirement for several micrograms of high-quality genomic DNA in order to obtain reliable DNA fingerprints. Because forensic specimens are often old and yield small quantities of degraded DNA, MLPs are not generally suitable for forensic analysis, although they were successfully implemented in a number of early criminal investigations (Gill and Werrett, 1987).

DNA profiling using single-locus probes (SLPs)

To circumvent the limitations of MLPs, specific cloned minisatellites were used as SLPs to produce simpler 'DNA profiles' and were applied in criminal casework even before MLPs were commercially established as the standard method for paternity testing (Figure 5.2). Since each SLP detects only a single minisatel-

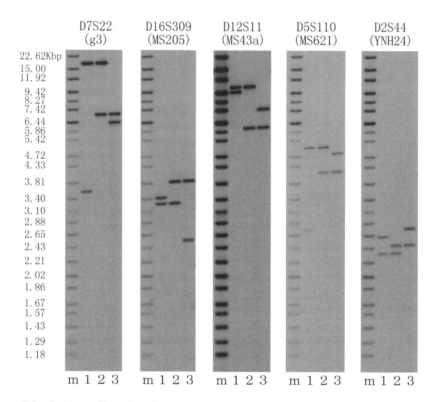

Figure 5.2 DNA profile using single-locus probes: results from a paternity test exploring five loci. Samples were obtained from the mother (lane 1), child (lane 2) and alleged father (lane 3). The numbers at the top of each autorad represent the name of the loci examined. Size marker is indicated by 'm'. Paternity was successfully established

lite, it produces two band (two allele) patterns, but is still highly polymorphic due to the use of hypervariable minisatellites. Single-locus probes have considerable advantages over MLPs in analysing forensic specimens. The method is far more sensitive, with the limit of band detection at around 10 ng of genomic DNA. Mixed DNA samples such as semen in vaginal swabs can be analysed with relative ease because the original DNA (from the victim) has only two bands and subsequent autoradiography will reveal whether the profile has more than two bands. Secondly, because allele sizes can be estimated and included in databases, the comparison of samples does not require side-by-side electrophoresis, which overcomes the inter-blot comparison problem associated with DNA fingerprinting. However, owing to continuous allele size distribution and the resolution limits of agarose gel electrophoresis, SLP allele sizing cannot be done with absolute precision. Two procedures for determining allele frequencies and match probabilities in the face of measurement errors are the floating-bin and the fixed-bin methods (Herrin, 1992). The true discriminating power of SLPs was compromised for no appropriate reason by genetically inappropriate calculations based on the 'ceiling principle' or 'interim ceiling principle' invented by the National Research Council of the United States (National Research Council, 1992); these calculations were abandoned in 1996 (National Research Council, 1996).

Amplified fragment length polymorphisms (AFLPs)

Some minisatellite loci with relatively short (~1000 bp) alleles can be amplified via polymerase chain reaction (PCR) to yield AFLPs. Before PCR-based typing of microsatellite loci was established, one such minisatellite locus called *D1S80* (Nakamura *et al.*, 1988; Kasai *et al.*, 1990) was used extensively within the realm of forensic DNA analysis. (Figure 5.3a). In the *D1S80* system, fragments between the range of 14–42 repeat units (16 bp per repeat) are amplified to yield alleles considerably smaller than the fragments normally analysed in DNA fingerprinting and profiling.

In contrast to RFLP analysis, *D1S80* alleles fall into discrete size classes and thus can be compared directly to a standard composite of most alleles (an allelic ladder) on the same gel (see Figure 5.3a). This was a significant improvement that was also employed in subsequent STR systems. In *D1S80*, the amplified DNA fragments were commonly detected using a silver stain, with the final profile usually showing two bands from which the numbers of repeat units can be easily estimated. Although the *D1S80* locus in particular contains four common alleles with frequencies of >10% in Japanese people, the likelihood of discrimination between two unrelated individuals is still high (0.977 in Japanese people; Member of TWGFDM, 1995).

This has been the preferred locus for analysing forensic specimens worldwide. In addition to this locus, PCR-based methods utilizing commercial kits that

(a)

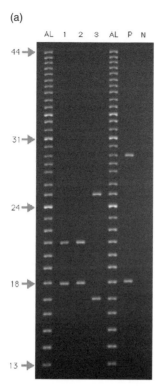

Figure 5.3 Detection of MCT118 (*D1S80*). (a) Agarose gel electrophoresis followed by silver-staining. Top figures indicate bloodstain (lane 1), victim (lane 2) and suspect (lane 3). The allelic ladder marker (AL) is displayed with the number of repeats indicated to the left. Positive and negative controls are expressed in lanes P and N, respectively. (b) Fluorescent labeled primer and capillary electrophoresis. Figures to the left indicate sample numbers, and AL, PC and NC represent the allelic ladder, positive control and negative control, respectively. The upper numbers below the peaks in each sample indicate the number of repeats and the lower numbers indicate the peak intensity

exploit single nucleotide polymorphisms (SNPs), including *HLA-DQα* and the five PolyMarker loci (*LDLR, GYPA, HBGG, D7S8* and *Gc*), were popularly used in the 1990s. Recently, the National Institute of Police Science (Japan) developed a novel genotyping method that involves PCR and fluorescent-labelled primers detected by capillary electrophoresis (Figure 5.3b). This method has allowed for higher accuracy since the results are not prone to human error.

However, after the detailed evaluation of allele frequency data in the world population at *D1S80* (Peterson *et al.*, 2000) and increasing practical usage of multiplex STR systems, *D1S80* has been gradually abandoned in favour of STRs.

(b)

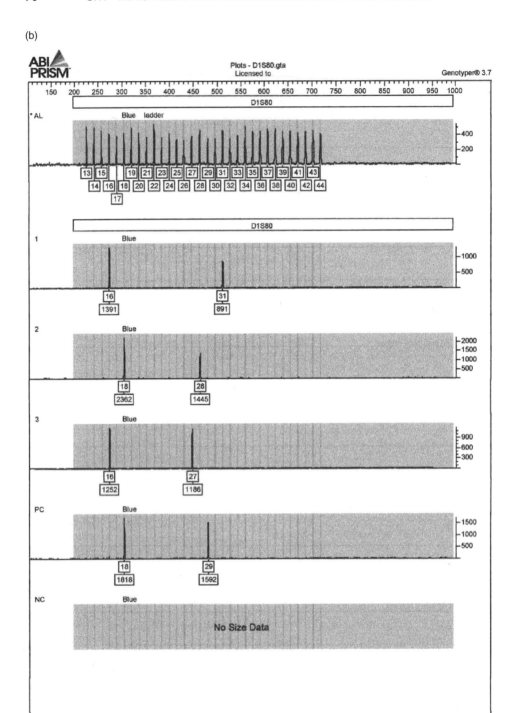

Figure 5.3   (b) *Continued*

Minisatellite variable repeat (MVR)-PCR

The majority of minisatellite loci consist of heterogeneous arrays of two or more repeat unit types (MVRs) that differ slightly. The human hypervariable minisatellites characterized to date vary not only in the number of repeat copies (allele length) but also in the interspersion pattern of variant repeat units within the array. This internal variation provides a highly informative approach to the study of allelic variation and the processes of mutation. Interspersion patterns can be determined by MVR mapping and this reveals far more variability than allele length analysis.

The first MVR mapping was developed at locus *D1S8* (minisatellite MS32) (Wong *et al.*, 1987). This particular minisatellite consists of a 29 bp repeat unit showing two classes of MVR that differ by a single base substitution, resulting in the presence or absence of an *Hae*III restriction site (Jeffreys *et al.*, 1990). Extremely high levels of variation in the interspersion patterns of two types of repeat within the alleles have been revealed by *Hae*III digestion of PCR-amplified alleles. Subsequently, a technically simpler version of a PCR-based mapping system (MVR-PCR) was invented (Jeffreys *et al.*, 1991b) (see also a detailed review by Jeffreys *et al.*, 1993). In addition to a primer at a fixed site in the DNA flanking the minisatellite, this assay reveals the internal variation of repeats by using an MVR primer specific to one or more types of variant repeat. The MVR-specific primers at low concentration will bind to just one of their complementary repeat units in the PCR, thus MVR locations will be represented as an array of sequentially sized, amplified DNA fragments. To analyse two different types of repeat, MVR-PCR uses a single flanking primer plus two different MVR primers to generate two complementary ladders of amplified products corresponding to the length between the flanking primer and the location of one or other of the repeat types within the minisatellite repeat array. In order to prevent the progressive shortening of PCR products by internal priming of the MVR-specific primers, 'tagged' amplification that uncouples MVR detection from subsequent amplification is used.

The MVR-PCR products are separated by electrophoresis on an agarose gel and detected by Southern blotting and hybridization to an isotope-labelled minisatellite probe. The PCR reaction for each of the two MVR-specific primers is carried out in a separate tube, and products are loaded in adjacent lanes in an agarose gel. Some work has been done that involves the labelling of variant primers with different-coloured fluorescent tags and amplifying both sets of MVR products in a single reaction, to facilitate comparison of the complementary MVR profiles (Rodriguez-Calvo *et al.*, 1996; Hau and Watson, 2000). The MVR-PCR technique has revealed enormous levels of allelic variation at several human hypervariable minisatellites: MS32 (*D1S8*) (Jeffreys *et al.*, 1991b; Tamaki *et al.*, 1992a, 1992b, 1993), MS31A (*D7S21*) (Neil and Jeffreys, 1993; Huang *et al.*, 1996), MS205 (*D16S309*) (Armour *et al.*, 1993; May *et al.*, 1996), CEB1 (*D2S90*) (Buard and Vergnaud, 1994), g3 (*D7S22*) (Andreassen and Olaisen,

1998), YNH24 (*D2S44*) (Holmlund and Lindblom, 1998), B6.7 (Tamaki *et al.*, 1999; Mizukoshi *et al.*, 2002) and the insulin minisatellite (Stead and Jeffreys, 2000).

At a minority of loci such as MS32, where almost all repeat units are of the same length, a diploid MVR map of the interspersion patterns of repeats from two alleles superimposed can be generated from total genomic DNA and encoded as a digital diploid code (Jeffreys *et al.*, 1991b). Since different length repeats will cause the MVR maps of each allele to drift out of register, most minisatellites are not amenable to this diploid coding, and instead only single allele coding is possible. Both single allele and diploid codes are highly suitable for computer databasing and analysis.

Allele-specific MVR-PCR methods (Monckton *et al.*, 1993) have been developed to map single alleles from total genomic DNA using allele-specific PCR primers directed to polymorphic SNP sites in the DNA flanking the minisatellite (Figure 5.4). These SNP primer pairs are identical, with the exception of a 3′ terminal mismatch that corresponds to the variant flanking base. This method is extremely expedient since it does not involve the time-consuming separation of the alleles by agarose gel electrophoresis. Another added advantage of allele-

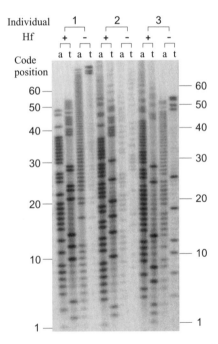

Figure 5.4   Allele-specific MVR-PCR of MS32 (*D1S8*). Alleles from the genomic DNA of three individuals were directly mapped utilizing an SNP (Hf) flanking the minisatellites (allele-specific MVR-PCR). The numbers to both sides of the figure indicate the code position, e.g. in sample 1 the Hf+ allele can be read as 'atataaaaat . . .'

specific MVR-PCR is its ability to recover individual-specific typing data from DNA mixtures (Tamaki *et al.*, 1995a).

The MVR-PCR technique is the best approach for exploiting the potential of hypervariable minisatellite loci because of the unambiguous nature of MVR mapping and the generation of digital MVR codes suitable for computer analysis. Code generation does not require standardization of electrophoretic systems, is immune to gel distortions and band shifts, does not involve error-prone DNA fragment length measurement and does not require side-by-side comparisons of DNA samples on the same gel.

For these reasons, MVR-PCR is well-suited for use in forensic analysis. The potential for forensic applications have been demonstrated by obtaining authentic diploid MVR coding ladders from only 1 ng of genomic DNA from bloodstains, saliva stains, seminal stains and plucked hair roots (Yamamoto *et al.*, 1994a). This is done by determining the source of saliva on a used postage stamp (Hopkins *et al.*, 1994), by making MVR coding ladders quickly without any need for blotting and hybridization (Yamamoto *et al.*, 1994b) and by maternal identification from remains of an infant and placenta (Tamaki *et al.*, 1995a). While forensic samples in general contain partially degraded DNA, MVR-PCR does not require intact minisatellite alleles. Such DNA samples yield truncated codes due to the disappearance of longer PCR products, but these codes are still compatible with the original allele information. Although replicate runs on the same sample and reading consensus codes are required, reliable codes can be obtained from as low as 100 pg of genomic DNA by MVR-PCR at MS32 (Jeffreys *et al.*, 1991b).

The MVR-PCR method at MS32 and at minisatellite MS31A has also been applied to paternity testing. The potential for establishing paternity in cases lacking a mother was demonstrated by a major contribution to the paternity index made possible by the extremely rare paternal alleles at these two loci (Huang *et al.*, 1999). Similarly, these rare alleles proved vital in confirming the relationship between a boy and his alleged grandparents despite an inconsistency (i.e. mutant allele) between the father and grandparents at one of the STR loci (Yamamoto *et al.*, 2001). Significant germline mutation rates to new length alleles have been observed at some hypervariable loci (Jeffreys *et al.*, 1988), which will generate false paternal exclusions in about 1.8% of paternity cases. In such cases, allele length measurements do not allow the distinction of non-paternity from mutation. In contrast, detailed knowledge of the mutation processes, coupled with MVR analysis of allele structures, can help distinguish mutations from non-paternity. This theory was tested at MS32 using both real and simulated allele data (Tamaki *et al.*, 2000). Since MVR-PCR allows information to be recovered from at least 40 repeat units, a mutant paternal allele will be identical to the progenitor paternal allele – with the exception of the first few repeats. Most germline mutation events altering repeat array structures are targeted to this region, most likely due to its proximity to a flanking recombination hotspot that appears to drive repeat instability (reviewed in Jeffreys *et al.*, 1999). Thus, a mutant paternal allele in a child will tend to resemble one of the father's alleles more than most other alleles in the population. This approach is unlikely to work at extremely hypervariable minisatellite loci such as CEB1 and B6.7, given their very high rate of germline instability coupled with complex germline mutation events that can radically alter allele structure in a single mutation event (Buard *et al.*, 1998; Tamaki *et al.*, 1999).

The MVR-PCR technique reveals enormous levels of variation, unmatched by any other single-locus typing system. At MS32, for example, almost all alleles in several ethnic populations surveyed were different. However, different alleles can show significant similarities in repeat organization (Jeffreys *et al.*, 1991b). Heuristic dot-matrix algorithms have been developed to identify significant allele alignments and have shown that approximately three-quarters of alleles mapped to date can be grouped into over 100 sets of alignable alleles, indicating relatively ancient groups of related alleles present in diverse populations (Tamaki *et al.*, 1995b). Some small groups of alleles can display a strong tendency to be population-specific, consistent with recent divergence from a common ancestral allele (Figure 5.5). In most groups, the 5′ ends of the aligned MVR maps show most variability due to the existence of the flanking recombination hotspot, therefore MVR allele analysis at MS32 can serve not only as a tool for individual identification but also for giving clues regarding the ethnic background of an individual (Tamaki *et al.*, 1995b). Another locus that provided a clear and detailed view of allelic divergence between African and non-African populations is MS205 (Armour *et al.*, 1996). A restricted set of allele families was found in non-African populations and formed a subset of the much greater diversity seen in Africans, which supports arguments for a recent African origin for modern human diversity at this locus. Very similar findings emerged from MVR analyses of the insulin minisatellite, again pointing to a major bottleneck in the 'Out of Africa' founding of non-African populations (Stead and Jeffreys, 2002).

Finally, MVR-PCR has been developed at the Y-chromosome-specific variable minisatellite *DYF155S1* (MSY1), with potential for extracting male-specific information from mixed male/female samples (Jobling *et al.*, 1998). This marker is also useful for paternity exclusion and, if adequate population data are available and the allele is rare, can be used in individual identification and paternity inclusion. The now-extensive use of Y-chromosomal microsatellites in forensic applications is discussed in Chapter 9.

Unfortunately, MVR-PCR has rarely been used in forensic analysis despite its simplicity, considerable discriminatory power and its ability to reveal enormous levels of variation.

5.3 Microsatellites

As with minisatellites, microsatellites are tandemly repeated DNA sequences and are also known as simple sequence repeats or STRs. They consist of repeat

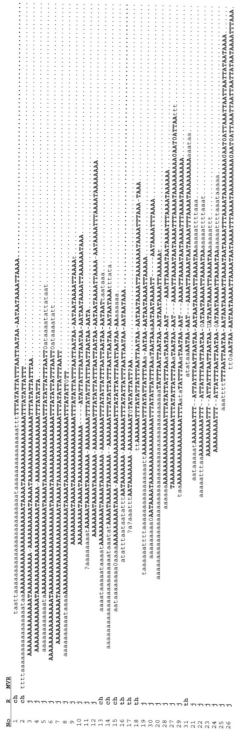

Figure 5.5 An example of alignable MS32 alleles. The ethnic origin (R: j, Japanese; ch, Chinese; th, Thai) and MVR code of a- and t-type repeats are shown for each allele. Haplotypic MVR map segments shared by related alleles are shown in uppercase and divergences by lowercase. Gaps (-) have been introduced to improve alignments. Some alleles show uncertain positions (?) and the unknown haplotypes of long alleles beyond the mapped region are indicated by (...)

units of 1–5 bp typically repeated 5–30 times. Most microsatellite loci can be efficiently amplified by standard PCR since the repeat regions are shorter than 100 bp. Some microsatellites can display substantial polymorphism – though far less than at the most variable minisatellites – and are found abundantly through-out the human genome, which contains over a million of these loci (International Human Genome Sequencing Consortium, 2001). Microsatellites are particularly suitable for analysing forensic specimens containing degraded and/or limited amounts of DNA. The first forensic application of microsatellites involved typing the skeletal remains of a murder victim (Hagelberg *et al.*, 1991), followed by the identification of Josef Mengele, the Auschwitz 'Angel of Death' (Jeffreys *et al.*, 1992). Although the spurious shadow or stutter bands often observed at dinucleotide repeat loci can make interpretation difficult, small PCR products can be sized with precision by polyacrylamide gel electrophoresis (PAGE). Because interpretation can sometimes be difficult, current typing systems employ the use of microsatellites with repeat units 4 bp long to minimize problems of stuttering. The microsatellite approach allows very high throughput via multi-plex PCR (single-tube PCR reactions that amplify multiple loci) and fluorescent detection systems have been developed to allow substantial automation of gel or capillary electrophoresis and DNA profile interpretation (Edwards *et al.*, 1991).

Microsatellite analysis is highly sensitive and can even recover information from a single cell. Discrete allele sizes, combined with automated electrophoresis and interpretation, make unambiguous assignment of alleles possible. For these reasons, the USA, most European nations and numerous other countries have established ever-growing databases. In the United Kingdom, 10 autosomal STR loci plus the amelogenin sex test, typed using the 'second-generation multiplex' (SGM) Plus system, are used in forensic practice (Table 5.1; Tamaki and Jeffreys, 2005). The SGM Plus loci generate random match probabilities of typically 10^{-11} between two unrelated individuals in the three UK racial groups (Caucasian, Afro-Caribbean and Asian) (Foreman and Evett, 2001). The US FBI CODIS (Combined DNA Index System) uses 13 STRs plus amelogenin. These loci produce extremely low random match probabilities. For example, around 60% of Japanese individuals show match probabilities of 10^{-14}–10^{-17} (estimates from Yoshida *et al.*, 2003, and Hayashida *et al.*, 2003). In 2003, the Japanese National Police Agency (NPA) introduced a nine-microsatellite locus system, which uses some CODIS loci (the AmpFlSTR® Profiler® kit), for analysing forensic specimens. From allele frequencies at these nine loci (Yoshida *et al.*, 2003), around 60% of Japanese individuals have match probabilities of 10^{-9}–10^{-11}. This particular system also detects X–Y homologous amelogenin genes to reveal the sex of a sample. Recently, new multiplexes that amplify 16 loci in a single reaction, including amelogenin, have been commercially developed (Figure 5.6). These systems produce even lower match probabilities without sacrificing sensitivity, and it is likely that such systems will remain the standard for analys-ing specimens in both forensic applications and paternity tests. The Japanese

Table 5.1 Properties of STR loci used in SGM Plus and FBI CODIS (Tamaki and Jeffreys, 2005)

Locus	SGM Plus™	FBI CODIS	Chromosome	Genomic position (kb)	Repeat motif	No. alleles in Japanese[a]	Power of discrimination in Japanese
TPOX	–	+	2	1 470	AATG	7	0.814
D2S1338	+	–	2	218 705	TTCC	14	0.971
D3S1358	+	+	3	45 557	TCT (A or G)	10	0.852
HUMFIBRA/FGA	+	+	4	155 865	CTTT	17	0.962
D5S818	–	+	5	123 139	AGAT	9	0.919
CSF1PO	–	+	5	149 436	AGAT	10	0.886
D7S820	–	+	7	82 751	AGAT	12	0.908
D8S1179	+	+	8	125 976	TCTA	9	0.952
HUMTH01	+	+	11	2 149	AATG	7	0.876
HUMVWA	+	+	12	5 963	AGAT	10	0.921
D13S317	–	+	13	81 620	AGAT	9	0.934
D16S539	+	+	16	84 944	AGAT	8	0.908
D18S51	+	+	18	59 100	AGAA	17	0.965
D19S433	+	–	19	35 109	AAGG	13	0.907
D21S11	+	+	21	19 476	TCTA	14	0.928

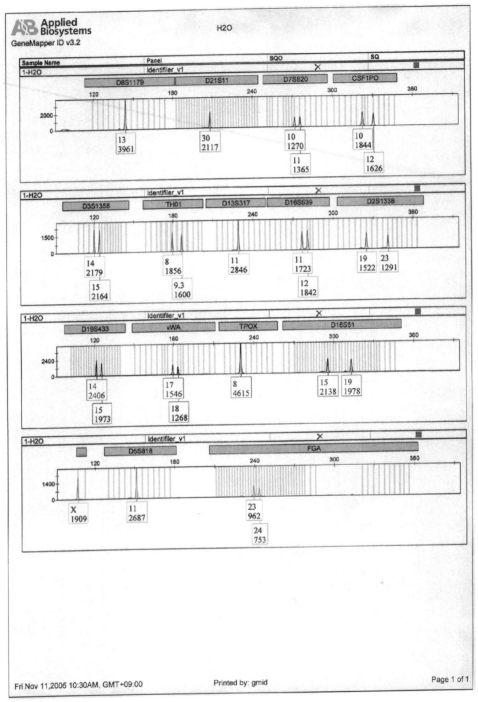

Figure 5.6   Genotyping at 15 microsatellite (STR) loci via multiplex PCR and capillary electrophoresis using the AmpFISTR® identifier kit. The genotype was determined by GeneMapper® software. The upper set of numbers located beneath the peaks indicate the number of alleles and the lower set indicate the peak intensity

NPA has taken initiative in implementing this system in its forensic applications since 2006.

In situations where a direct comparison between evidence and a suspect is being made, mutation rates are an insignificant issue. However, with comparisons between relatives in parentage testing and kinship analysis – such as victim identification in mass disasters – mutational events may lead to false negatives (Biesecker *et al.*, 2005). In comparison to minisatellites, the mutation rate in most microsatellites is much lower, 0.2% at the most. Consequently, microsatellites are better suited for paternity testing as long as a greater number of loci are used compared with minisatellites, in order to compensate for the relatively low level of polymorphism at each locus.

In recent years the development of commercialized kits has reduced what once was a painstaking task into a simple procedure. Since its inception, these kits have achieved new heights in quality and the microsatellite systems commercially available today have a high standard of sensitivity. Unfortunately, the forensic specimens that we encounter in the field are often degraded or fragmented. As a result, when amplification of DNA is attempted at loci yielding relatively long PCR products (e.g. CSF1PO, D2S1338, D18551 and FGA in Figure 5.6), we often fail to retrieve any PCR products, making it impossible to determine the genotypes of these particular loci. To remedy this situation, innovative PCR techniques known as 'mini-STRs' are currently being developed (Butler *et al.*, 2003), which involve bringing primer annealing sites closer to the microsatellites in order to shorten the PCR products, and selecting loci with relatively short allele length distributions,. Recently, two different parallel strategies have emerged in Europe. The first employs a 13-STR loci multiplex incorporating three mini-STRs into the currently used multiplex test. The second strategy employs a multiplex of six high-molecular-weight STRs (also in current use), modified to provide smaller amplicons with an additional two loci of high discriminating power. Eventually, the two strategies will merge to provide a single multiplex of 15 STR loci (Gill *et al.*, 2006). In this section, we have addressed the applications of autosomal microsatellites; however, microsatellites in the Y chromosome (YSTRs) are of equal importance and should not be ignored (see Chapter 9). Detailed information on forensic microsatellites can be obtained at STRBase (Ruitberg *et al.*, 2001).

In the future, technological advances, along with newer and more improved methods, may give rise to an alternative typing platform utilizing SNPs. This could greatly accelerate typing and offers the potential for DNA analysis to be conducted at the scene of crime. However, even with the introduction of an alternative typing platform, the multiplex microsatellite markers employed today will remain in use for the time being, primarily because millions of DNA profiles – such as those contained in national databases – were generated with STR loci and will require ongoing maintenance.

5.4 Acknowledgements

The author wishes to thank and acknowledge the following parties for their invaluable contributions in the formulation of this chapter. First, to SRL Inc. for the generous provision of figures on DNA fingerprints (Figure 5.1) and DNA profiles (Figure 5.2). Thanks also go to Dr Ken Kasai from NIPS for his general guidance and for providing Figure 5.3, and to Professor Alec Jeffreys for his invaluable input and for providing data on MVR haplotypes.

5.5 References

Andreassen, R. and Olaisen, B. (1998) De novo mutations and allelic diversity at mini-satellite locus D7S22 investigated by allele-specific four-state MVR-PCR analysis. *Hum. Mol. Genet.* 7: 2113–2120.

Armour, J.A., Anttinen, T., May, C.A., Vega, E.E., Sajantila, A., Kidd, J.R., Kidd, K.K., Bertranpetit, J., Paabo, S. and Jeffreys, A.J. (1996) Minisatellite diversity supports a recent African origin for modern humans. *Nat. Genet.* 13: 154–160.

Armour, J.A., Harris, P.C. and Jeffreys, A.J. (1993) Allelic diversity at minisatellite MS205 (D168309): evidence for polarized variability. *Hum. Mol. Genet.* 2: 1137–1145.

Biesecker, L.G., Bailey-Wilson, J.E., Ballantyne, J., Baum, H., Bieber, F.R., Brenner, C., et al. (2005) Epidemiology. DNA identifications after the 9/11 World Trade Center attack. *Science* 310: 1122–1123.

Buard, J., Bourdet, A., Yardley, J., Dubrova, Y. and Jeffreys, A.J. (1998) Influences of array size and homogeneity on minisatellite mutation. *EMBO J.* 17: 3495–3502.

Buard, J. and Vergnaud, G. (1994) Complex recombination events at the hypermutable minisatellite CEBl (D2S90). *EMBO J.* 13: 3203–3210.

Butler, J.M., Shen, Y. and McCord, B.R. (2003) The development of reduced size STR amplicons as tools for analysis of degraded DNA. *J. Forensic Sci.* 48: 1054–1064.

Edwards, A., Civitello, A., Hammond, H.A. and Caskey, C.T. (1991) DNA typing and genetic mapping with trimeric and tetrameric tandem repeats. *Am. J. Hum. Genet.* 49: 746–756.

Foreman, L.A. and Evett, I.W. (2001) Statistical analyses to support forensic inter-pretation for a new ten-locus STR profiling system. *Int. J. Legal Med.* 114: 147–155.

Gill, P., Fereday, L., Morling, N. and Schneider, P.M. (2006) New multiplexes for Europe – amendments and clarification of strategic development. *Forensic Sci. Int. Jan.* 163: 155–157.

Gill, P. and Werrett, D.J. (1987) Exclusion of a man charged with murder by DNA fin-gerprinting. *Forensic Sci. Int.* 35: 145–148.

Hagelberg, E., Gray, I.C. and Jeffreys, A.J. (1991) Identification of the skeletal remains of a murder victim by DNA analysis. *Nature* 352: 427–429.

Hau, P. and Watson, N. (2000) Sequencing and four-state minisatellite variant repeat mapping of the D1S7 locus (MS1) by fluorescence detection. *Electrophoresis* 21: 1478–1483.

Hayashida, M., Itakura, Y., Nagashima, T., Nata, M. and Funayama, M. (2003) Polymorphism of 17 STRs by multiplex analysis in Japanese population. *Forensic Sci. Int.* **133**: 250–253.

Herrin, G., Jr (1992) A comparison of models used for calculation of RFLP pattern frequencies. *J. Forensic Sci.* **37**: 1640–1651.

Holmlund, G. and Lindblom, B. (1998) Different ancestor alleles: a reason for the bimodal fragment size distribution in the minisatellite D2S44 (YNH24). *Eur. J. Hum. Genet.* **6**: 597–602.

Hopkins, B., Williams, N.J., Webb, M.B., Debenham, P.C. and Jeffreys, A.J. (1994) The use of minisatellite variant repeat-polymerase chain reaction (MVR-PCR) to determine the source of saliva on a used postage stamp. *J. Forensic Sci.* **39**: 526–531.

Huang, X.L., Tamaki, K., Yamamoto, T., Suzuki, K., Nozawa, H., Uchihi, R., Katsumata, Y. and Neil, D.L. (1996) Analysis of allelic structures at the D7S21 (MS31A) locus in the Japanese, using minisatellite variant repeat mapping by PCR (MVR-PCR). *Ann. Hum. Genet.* **60**: 271–279.

Huang, X.L., Tamaki, K., Yamamoto, T., Yoshimoto, T., Mizutani, M., Leong, Y.K., Tanaka, M., Nozawa, H., Uchihi, R. and Katsumata, Y. (1999) Evaluation of the paternity probability on an application of minisatellite variant repeat mapping using polymerase chain reaction (MVR-PCR) to paternity testing. *Leg. Med.* **1**: 37–43.

International Human Genome Sequencing Consortium (2001) Initial sequencing and analysis of the human genome. *Nature* **409**: 860–921.

Jeffreys, A.J., Allen, M.J., Hagelberg, E. and Sonnberg, A. (1992) Identification of the skeletal remains of Josef Mengele by DNA analysis. *Forensic Sci. Int.* **56**: 65–76.

Jeffreys, A.J., Barber, R., Bois, P., Buard, J., Dubrova, Y.E., Grant, G., Hollies, C.R., May, C.A., Neumann, R., Panayi, M., Ritchie, A.E., Shone, A.C., Signer, E., Stead, J.D. and Tamaki, K. (1999) Human minisatellites, repeat DNA instability and meiotic recombination. *Electrophoresis* **20**: 1665–1675.

Jeffreys, A.J., Brookfield, J.F. and Semeonoff, R. (1985c) Positive identification of an immigration test-case using human DNA fingerprints. *Nature* **317**: 818–819.

Jeffreys, A.J., MacLeod, A., Tamaki, K., Neil, D.L. and Monckton, D.G. (1991b) Minisatellite repeat coding as a digital approach to DNA typing. *Nature* **354**: 204–209.

Jeffreys, A.J., Monckton, D.G., Tamaki, K., Neil, D.L., Armour, J.A.L., MacLeod, A., Collick, A., Allen, M. and Jobling, A. (1993) Minisatellite variant repeat mapping: application to DNA typing and mutation analysis. In: *DNA fingerprinting, State of the Science* (S.D.J. Pena, *et al.*, eds), Birkhäuser-Verlag, Basel, pp. 125–139.

Jeffreys, A.J., Neumann, R. and Wilson, V. (1990) Repeat unit sequence variation in minisatellites: a novel source of DNA polymorphism for studying variation and mutation by single molecule analysis. *Cell* **60**: 473–485.

Jeffreys, A.J., Royle, N.J., Wilson, V. and Wong, Z. (1988) Spontaneous mutation rates to new length alleles at tandem-repetitive hypervariable loci in human DNA. *Nature* **332**: 278–281.

Jeffreys, A.J., Turner, M. and Debenham, P. (1991a) The efficiency of multilocus DNA fingerprint probes for individualization and establishment of family relationships, determined from extensive casework. *Am. J. Hum. Genet.* **48**: 824–840.

Jeffreys, A.J., Wilson, V. and Thein, S.L. (1985a) Hypervariable 'minisatellite' regions in human DNA. *Nature* **314**: 67–73.

Jeffreys, A.J., Wilson, V. and Thein, S.L. (1985b) Individual-specific 'fingerprints' of human DNA. *Nature* **316**: 76–79.

Jobling, M.A., Bouzekri, N. and Taylor, P.G. (1998) Hypervariable digital DNA codes for human paternal lineages: MVR-PCR at the Y-specific minisatellite, MSY1 (DYF155S1). *Hum. Mol. Genet.* **7**: 643–653.

Kasai, K., Nakamura, Y. and White, R. (1990) Amplification of a variable number of tandem repeats (VNTR) locus (pMCT118) by the polymerase chain reaction (PCR) and its application to forensic science. *J. Forensic Sci.* **35**: 1196–1200.

May, C.A., Jeffreys, A.J., and Armour, J.A. (1996) Mutation rate heterogeneity and the generation of allele diversity at the human minisatellite MS205 (D16S309). *Hum. Mol. Genet.* **5**: 1823–1833.

Member of TWGFDM (1995) Reassessment on frequency distribution of MCT118 and HLADQα alleles and genotypes in Japanese population along with the comparison with other population data (in Japanese). *Kakeiken Houkoku* **48**: 171–185.

Mizukoshi, T., Tamaki, K., Azumi, J., Matsumoto, H., Imai, K. and Jeffreys, A.J. (2002) Allelic structures at hypervariable minisatellite B6.7 in Japanese show population specificity. *J. Hum. Genet.* **47**: 232–238.

Monckton, D.G., Tamaki, K., MacLeod, A., Neil, D.L. and Jeffreys, A.J. (1993) Allele-specific MVR-PCR analysis at minisatellite DlS8. *Hum. Mol. Genet.* **2**: 513–519.

Nakamura, Y., Carlson, M., Krapcho, K. and White, R. (1988) Isolation and mapping of a polymorphic DNA sequence (pMCT118) on chromosome 1p [D1S80]. *Nucleic Acids Res.* **16**: 9364.

National Research Council (1992) DNA Technology in Forensic Science, National Academy Press, Washington, DC.

National Research Council (1996) The Evaluation of Forensic DNA Evidence, National Academy Press, Washington, DC.

Neil, D.L. and Jeffreys, A.J. (1993) Digital DNA typing at a second hypervariable locus by minisatellite variant repeat mapping. *Hum. Mol. Genet.* **2**: 1129–1135.

Peterson, B.L., Su, B., Chakraborty, R., Budowle, B. and Gaensslen, R.E. (2000) World population data for the HLA-DQA1, PM and D1S80 loci with least and most common profile frequencies for combinations of loci estimated following NRC II guidelines. *J. Forensic Sci.* **45**: 118–146.

Rodriguez-Calvo, M.S., Bellas, S., Soto, J.L., Barros, F. and Carracedo, A. (1996) Comparison of different electrophoretic methods for digital typing of the MS32 (D1S8) locus. *Electrophoresis.* **17**: 1294–1298.

Royle, N.J., Clarkson, R.E., Wong, Z. and Jeffreys, A.J. (1988) Clustering of hypervariable minisatellites in the proterminal regions of human autosomes. *Genomics* **3**: 352–360.

Ruitberg, C.M., Reeder, D.J. and Butler, J.M. (2001) STRBase: a short tandem repeat DNA database for the human identity testing community. *Nucleic Acids Res.* **29**: 320–322.

Stead, J.D. and Jeffreys, A.J. (2000) Allele diversity and germline mutation at the insulin minisatellite. *Hum. Mol. Genet.* **9**: 713–723.

Stead, J.D. and Jeffreys, A.J. (2002) Structural analysis of insulin minisatellite alleles reveals unusually large differences in diversity between Africans and non-Africans. *Am. J. Hum. Genet.* **71**: 1273–1284.

Tamaki, K., Brenner, C.H. and Jeffreys, A.J. (2000) Distinguishing minisatellite mutation from non-paternity by MVR-PCR. *Forensic Sci. Int.* **113**: 55–62.

Tamaki, K., Huang, X.L., Yamamoto, T., Uchihi, R., Nozawa, H. and Katsumata, Y. (1995a) Applications of minisatellite variant repeat (MVR) mapping for maternal identification from remains of an infant and placenta. *J. Forensic Sci.* **40**: 695–700.

Tamaki, K., Huang, X.L., Yamamoto, T., Uchihi, R., Nozawa, H., Katsumata, Y. and Jeffreys, A.J. (1995b) Characterisation of MS32 alleles in the Japanese population by MVR-PCR analysis (in Japanese). In: *DNA Polymorphism*, vol. 3 (A. Sawaguchi and S. Nakamura, eds), Toyo Shoten, Tokyo, pp. 137–143.

Tamaki, K. and Jeffreys, A.J. (2005) Human tandem repeat sequences in forensic DNA typing. *Leg. Med.* **7**: 244–250.

Tamaki, K., May, C.A., Dubrova, Y.E. and Jeffreys, A.J. (1999) Extremely complex repeat shuffling during germline mutation at human minisatellite B6.7. *Hum. Mol. Genet.* **8**: 879–888.

Tamaki, K., Monckton, D.G., MacLeod, A., Allen, M. and Jeffreys, A.J. (1993) Four-state MVR-PCR: increased discrimination of digital DNA typing by simultaneous analysis of two polymorphic sites within minisatellite variant repeats at D1S8. *Hum. Mol. Genet.* **2**: 1629–1632.

Tamaki, K., Monckton, D.G., MacLeod, A., Neil, D.L., Allen, M. and Jeffreys, A.J. (1992a) Minisatellite variant repeat (MVR) mapping: analysis of 'null' repeat units at D1S8. *Hum. Mol. Genet.* **1**: 401–406, 558.

Tamaki, K., Yamamoto, T., Uchihi, R., Kojima, T., Katsumata, Y. and Jeffreys, A.J. (1992b) DNA typing and analysis of the D1S8 (MS 32) allele in the Japanese population by the minisatellite variant repeat (MVR) mapping by polymerase chain reaction (PCR) assay. *Nippon Hoigaku Zasshi* **46**: 474–482.

Wong, Z., Wilson, V., Patel, I., Povey, S. and Jeffreys, A.J. (1987) Characterization of a panel of highly variable minisatellites cloned from human DNA. *Ann. Hum. Genet.* **51**: 269–288.

Yamamoto, T., Tamaki, K., Huang, X.L., Yoshimoto, T., Mizutani, M., Uchihi, R., Katsumata, Y. and Jeffreys, A.J. (2001) The application of minisatellite variant repeat mapping by PCR (MVR-PCR) in a paternity case showing false exclusion due to STR mutation. *J. Forensic Sci.* **46**: 374–378.

Yamamoto, T., Tamaki, K., Kojima, T., Uchihi, R. and Katsumata, Y. (1994a) Potential forensic applications of minisatellite variant repeat (MVR) mapping using the polymerase chain reaction (PCR) at D1S8. *J. Forensic Sci.* **39**: 743–750.

Yamamoto, T., Tamaki, K., Kojima, T., Uchihi, R., Katsumata, Y. and Jeffreys, A.J. (1994b) DNA typing of the D1S8 (MS32) locus by rapid detection minisatellite variant repeat (MVR) mapping using polymerase chain reaction (PCR) assay. *Forensic Sci. Int.* **66**: 69–75.

Yoshida, K., Mizuno, N., Fujii, K., Senju, H., Sekiguchi, K., Kasai, K. and Sato, H. (2003) Japanese population database for nine STR loci of the AmpFlSTR Profiler kit. *Forensic Sci. Int.* **132**: 166–167.

6

Application of SNPs in forensic casework

Claus Børsting, Juan J. Sanchez and Niels Morling

6.1 Introduction

Ever since the first human 'DNA fingerprint' was made by hybridization of multi-locus probes to variable number tandem repeats (VNTRs) (Jeffreys *et al.*, 1985) and the first DNA evidence was presented at court (Gill *et al.*, 1985), tandem repeat sequences have been the favoured targets for forensic DNA analyses. Today, short tandem repeats (STRs) with four nucleotide repeat units are preferred, mainly because they are easily amplified in polymerase chain reactions (PCRs) and because they are shorter than VNTRs and more likely to be intact in low-quality samples often recovered from crime scenes. Thirteen STRs located on different chromosomes were selected for the Combined DNA Indexing System (CODIS) database (Budowle *et al.*, 1999), and today there are several commercial products available that allow amplification of all CODIS loci in one multiplex PCR. These kits are validated and used by forensic laboratories all over the world, and they have facilitated the development of standardized databases, which have proved to be highly valuable tools for national and international law enforcement. Each CODIS locus has many different alleles because mutations happen frequently in tandem repeat sequences (on average 1 mutation per 300 generations), and therefore it is highly likely that different alleles are found in different individuals, which makes tandem repeat sequences very suitable for identification purposes.

Traditionally, the main purpose of forensic genetic investigations has been the identification of human remains found at crime scenes and mass disasters, or the determination of family relations where the available information is disputed or uncertain (e.g. in paternity or immigration cases). However, with the increase in the number of DNA profiles in national and international databases of criminal offenders, DNA profiling has become more and more important as an investigative tool for the police. This trend will most likely continue as more knowledge of the human genome is disclosed and new forensic genetic tools are developed. In this chapter, the present and potential applications of single nucleotide polymorphisms (SNPs) are discussed.

6.2   Single nucleotide polymorphisms

An SNP is formed by a point mutation, where one base pair is substituted by another base pair. Per definition, a genetic variation at a single base pair locus is not considered to be an SNP unless at least two alleles have frequencies of more than 1% in a large, random population. Thus, 'private alleles' identified in small selected populations (e.g. families) are considered to be mutations and not SNPs.

The vast majority of SNPs have only two alleles because the mutation rate at a particular base pair position in the genome is extremely low (on average 1 mutation per 100 million generations) and it is highly unlikely that two point mutations happen at the same position. For this reason, SNPs can be used to distinquish between populations and the geographical history of a given population can be mapped by identifying the distribution of a particular SNP allele among existing populations. Recently, the National Geographic Society, IBM and the Waitt family foundation initiated a worldwide survey of human populations with the purpose of mapping all major human migrations since modern humans left Africa approximately 60 000 years ago (*www.nationalgeographic. com*). The markers used in this study are Y chromosome SNPs and mitochondrial DNA (mtDNA) SNPs, which are particularly useful for mapping large human migrations (Jobling and Tyler-Smith, 2003; Pakendorf and Stoneking, 2005), because the Y chromosome is inherited from father to son and the mtDNA genome is inherited from mother to any offspring without recombination. Thus, a point mutation in the Y chromosome or the mtDNA creates a new male or female lineage, respectively, and the lineage remains distinct from all other lineages in future generations.

The human genome consists of approximately 3000 million base pairs (bp) and the most recent estimate of SNPs is 10 million (Lai, 2006), which gives an average of one SNP per 300 base pairs. The density of SNPs across the genome varies up to ten-fold (Sachidanandam *et al.*, 2001), because of variations in selection pressure, and local recombination and mutation rates (Reich *et al.*, 2002), and the majority of SNPs are located in repeat regions, which are notoriously difficult to analyse. Nevertheless, SNPs are the best choice for construction of a dense set of polymorphic markers that cover the whole genome. The marker set can be used for studying association between the markers and a particular human trait or disease (International HapMap Consortium, 2005). Once an association has been found, a more detailed analysis of the region surrounding the relevant marker(s) can be performed and the polymorphism(s) responsible for the human trait(s) or disease(s) may be identified.

The human genome consists of 30 000–35 000 genes, but the coding regions of these genes only comprise 1.1–1.5% of the genome. The SNPs located outside coding regions can influence gene expression if they are located in regulatory DNA sequences (Wang *et al.*, 2005), but the majority of SNPs probably have very little or no functional consequences for the organism.

An SNP located in a gene may have diverse effects on the cellular function of the protein encoded by the gene. If the SNP is located in the coding region of the gene, the different alleles may encode different proteins, because the trinu-cleotide sequence (the codon) that codes for one amino acid differs. For example (see Figure 6.1), the codon TGC can be changed to TGT, TGG or TGA by a point mutation in the third position of the codon. The original C allele will code for the amino acid cysteine. The T allele will also code for the amino acid cysteine (known as a silent mutation), whereas the G allele will code for the amino acid tryptophan (known as a missence mutation) and the A allele will code for termination of protein synthesis (known as a nonsense mutation). Obviously, the nonsense mutation is the most severe form of mutation and will almost always result in a non-functional protein. A missence mutation can have all kinds of consequences on the protein, including mis-folding, mis-placement and decreased or increased activity, which again may affect the organism in various ways. Even a silent mutation may not be neutral for the cell, because a silent mutation may affect the efficiency of protein synthesis and thus alter the cellular concentration of the protein.

DNA sequence	TTC GAC TG**C** AAA
Protein sequence	Phe Asp Cys Lys

DNA sequence	TTC GAC TG**T** AAA
Protein sequence	Phe Asp Cys Lys

DNA sequence	TTC GAC TG**G** AAA
Protein sequence	Phe Asp Trp Lys

DNA sequence	TTC GAC TG**A** AAA
Protein sequence	Phe Asp Stop

Figure 6.1. Examples of how an SNP may change the amino acid sequence of a protein

6.3   Single nucleotide polymorphism typing technology

Numerous methods have been developed for SNP genotyping (for review, see Kwok, 2001; Sobrino *et al.*, 2005). The methods all employ one of four common technologies: hybridization, primer extension, ligation or invasive cleavage. For forensic applications, single base extension (SBE) is currently the preferred method (Figure 6.2), because it is highly accurate and can be performed with the same instruments used for STR analyses (polymerase chain reaction machines and electrophoresis instruments). The SBE reaction is performed as consecutive cycles of denaturation of double-stranded DNA, annealing of the SBE primers to the polymerase chain reaction (PCR) products and single base extension. The SBE primer anneals to the single-stranded PCR product immediately upstream of the SNP position, and the DNA polymerase adds a fluorescently labelled dideoxyribonucleotide, complementary to the nucleotide in the SNP position, to the SBE primer. Single base extension reactions can be multiplexed and many SBE products

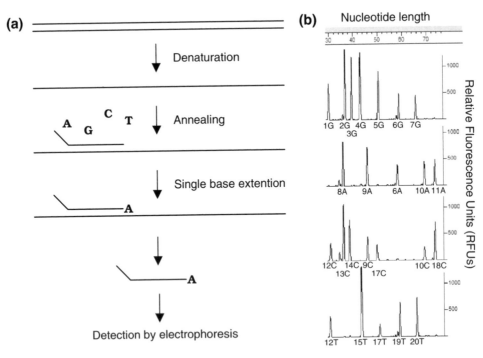

Figure 6.2.   (a) Schematic diagram of the SBE reaction. (b) Typical example of a multiplexed SBE reaction detected by electrophoresis. Twenty autosomal SNPs were typed based on the nucleotide lengths of the SBE primers and the fluorescent label of the added dideoxynucleotide. The tested individual was heterozygous in four positions (SNP numbers 9, 10, 12 and 17) and homozygous in 16 positions

can be analysed simultaneously by electrophoresis (Sanchez *et al.*, 2005). In the electropherogram, the length of the SBE primer identifies the SNP locus and the colour of the fluorescent label identifies the SNP allele (Figure 6.2b).

6.4 Single nucleotide polymorphisms for human identification

For human identification purposes, one SNP locus is obviously less informative than one STR locus, because the SNP locus has only two possible alleles and the STR locus typically has 8–15 different alleles. The match probability P for n SNP loci (between the SNP profiles from two randomly selected individuals) can be approximated by assuming that all SNPs are in Hardy–Weinberg equilibrium and that the frequency of the least common allele ρ is constant for all loci:

$$P = (\rho^2)^n + [2\rho(1-\rho)]^n + \left[(1-\rho)^2\right]^n \tag{6.1}$$

This is a simple function of ρ and n, and by comparing P to the match probabilities obtained for STRs it can be estimated that 50 SNPs give match probabilities equivalent to 12 STRs (Gill, 2001) if ρ is between 0.2 and 0.5; P has the highest value for $\rho = 0.5$, but P does not change very much when ρ is between 0.3 and 0.5. Thus, Equation (6.1) is a good estimate for the real match probability of a set of SNPs if the allele frequencies of the selected SNP loci are within this range, even though Equation (6.1) was calculated under the assumption that ρ is constant for all loci. In a paternity case, the power of exclusion, Z, can be calculated based on all possible genotype combinations of mother and child (Krawczak, 1999). For a biallelic SNP,

$$Z = \rho(1-\rho)[1-\rho(1-\rho)] \tag{6.2}$$

and for a multi-allelic STR,

$$Z = \sum_{i}^{m} \rho_i \left\{ (1-\rho_i)^2 + \rho_i(3\rho_i - 2)\left[\left(\sum_{i}^{m} \rho_i^2\right) - \rho_i^2\right] \right\} \tag{6.3}$$

where ρ_i is the frequency of the ith allele and m is the number of alleles.

Under the assumption that $\rho = 0.5$ for SNPs and $\rho_i = 1/m$ for STRs, 4–8 SNPs are needed to obtain the same power of exclusion as one STR. Using real Caucasian allele frequencies for 14 commonly used STRs, the average number of SNPs needed to obtain the same power of exclusion as one STR was 4.23 (SNPs with $\rho = 0.5$), 4.41 (SNPs with $\rho = 0.4$) or 5.04 (SNPs with $\rho = 0.3$), respectively (Krawczak, 1999). This indicates that 50–60 SNPs with $\rho = 0.3$–0.5 have the same discriminatory power for analysis of stains (match probability) and disputed family relations (power of exclusion) as the 13 CODIS loci currently in use by most forensic laboratories.

There are two important reasons why it is preferred to analyse 50–60 SNP loci instead of 13 STR loci. First of all, the length of the PCR product containing an SNP locus need only be the length of the PCR primers plus one base pair (the SNP position). In theory, a DNA sequence is unique if it is 16 bp long (4^{16} = 4295 million combinations, which is more than the number of base pairs in the human genome). Thus a PCR product containing an SNP locus need only be (2 × 16) + 1 = 33 bp long. In reality, the PCR primer design restrains the positioning of the PCR primer (Sanchez *et al.*, 2005), and the PCR product must be longer. Nevertheless, most SNPs can be amplified on PCR products less than 100 bp in length. In contrast, some CODIS STR alleles have up to 40 tandem repeat units, each with a size of 4 bp, and consequently the PCR products containing the STR locus need to be 200 bp or longer. In the commercial kits used in most forensic laboratories to amplify the 13 CODIS STRs, the loci are amplified in the same tube (multiplex PCR) and the lengths of the PCR products (i.e. the alleles) are determined by electrophoresis. In order to separate and identify the many PCR products, the longest PCR products have been designed to be 400–450 bp, but in highly degraded DNA samples the average length of DNA fragments is shorter than 150 bp, and therefore many of the STR loci are not amplified when the sample has been exposed to high temperatures or high humidity that degrades the DNA. So-called mini-STR kits have been developed (Coble and Butler, 2005; Asamura *et al.*, 2006; Opel *et al.*, 2006), where the PCR products have been reduced in length, but these kits only target 4–6 STRs and the discriminatory power is significantly smaller. Furthermore, unfinished extension products will be formed in higher numbers when PCR is performed on highly degraded DNA, because PCR primers may anneal to a strand where the target sequence is interrupted. The unfinished extension products from mini-STR kits consist almost exclusively of tandem repeat sequences and can anneal to many different positions in the STR target locus during subsequent PCR cycles. This will increase the risk of amplifying fragments with a different number of tandem repeats than was originally present in the sample. If such products are made during the first critical cycles of the PCR, false alleles may be detected and assigned to the sample. In contrast, unfinished extension products from amplification of SNP loci cannot create false SNP alleles, because they can only anneal to one unique position.

The second reason why SNPs are preferred over STRs is the low mutation rate of SNPs. If the investigated man and a child in a typical paternity case do not share any alleles for a given locus (genetic inconsistency), it indicates that the man is not the father, but one genetic inconsistency between the father and the child is observed in approximately 100(13 × 0.003) = 3.9% of the cases, where the 13 CODIS loci have been investigated. This is a highly unfortunate, but unavoidable, consequence of the relatively high mutation rate of tandem repeats. In rare cases where relatives (e.g. two brothers, or father and son) are investigated, it may even be impossible to draw a conclusion, because relatives share a high number of genetic markers and few genetic inconsistencies are expected between an investigated relative (e.g. uncle or grandfather), who is not the father, and a child. For comparison, one genetic inconsistency between the father and the child will be observed in approximately $100(60 \times 0.0000001) = 0.0006\%$ of the cases if 60 SNPs were investigated.

There are three reasons why STRs are preferred over SNPs. First, the national and international databases contain STR profiles from known criminal offenders, victims and samples collected from crime scenes during the past two decades. It is an overwhelming task to type all these samples again for SNP markers, and in some cases it is impossible, because the samples may have been used up and cannot be replaced. Secondly, samples collected from crime scenes often contain DNA from more than one person, and mixtures are difficult to detect when analysing DNA markers with only two alleles (Gill, 2001). In contrast, STRs are very useful for detection of mixtures, because two individuals are likely to have three or four different alleles in some STR systems. Sometimes, it is even possible to estimate the STR profiles of the two individuals based on the amplification strength of the alleles, and it is always possible to calculate a match probability between a reference sample from an individual and the mixture. If the mixture contains DNA from more than two people, SNPs will be almost useless, whereas STRs may still be used for calculation of match probabilities (Gill *et al.*, 2006). Thirdly, very little DNA is often recovered from crime scenes (e.g. from fingerprints or hair), and since all such samples are unique and cannot be replaced it is essential to obtain as much information as possible from every investigation performed on the sample. This is one of the reasons why amplification of multiple fragments (multiplex PCR) is important for forensic genetic analyses. In addition, multiplexing also simplifies analysis, reduces cost and decreases the number of times a sample is handled in the laboratory, which reduces the risk of contamination and mix-ups. However, it is difficult to develop PCR multiplexes with 5–10 fragments, and construction of a robust multiplex with 50–60 fragments is a very serious undertaking.

Nevertheless, several large SNP panels have been suggested for human identification purposes (Inagaki *et al.*, 2004; Dixon *et al.*, 2005; Kidd *et al.*, 2006), and recently a 52-SNP-plex assay was described (Sanchez *et al.*, 2006) by a group of five forensic laboratories (The SNP*for*ID consortium, *www.snpforid. org*), where 52 fragments are amplified in one multiplex PCR and the SNPs are detected by two multiplex SBE reactions (Figure 6.2). Each of the SNPs in the 52-SNP-plex assay maps to unique locations on the autosomal chromosomes and has a minimum distance of 100 kb from known genes. The SNPs are polymorphic in the three major population groups (Caucasian, African and Asian) with $\rho = 0.3$–0.5, and all SNPs can be amplified and detected from as little as 200 pg of genomic DNA (approximately the amount of genomic DNA in 30 human cells). Currently, this is the most promising SNP multiplex for human identification, and the European DNA Profiling Group (EDNAP), a working group under the International Society for Forensic Haemogenetics (ISFH), and commercial companies have expressed strong interest in the 52-SNP-plex.

6.5   Single nucleotide polymorphisms in mitochondrial DNA

Analysis of mitochondrial DNA (mtDNA) has been used successfully for a number of years in anthropology and forensic genetics (Budowle *et al.*, 2003). The high copy number of mitochondrial genomes (typically more than 3000) in a somatic cell makes it more likely to amplify mtDNA markers than nuclear DNA markers from old and highly degraded sample materials. In addition, mtDNA can also be collected from hair shafts, which do not contain any nuclear DNA, and hair shafts are one of the frequent samples recovered from scenes of violent crimes. Mitochondrial DNA is maternally inherited, and there is no recombination between mitochondrial genomes, because there is only one chromosome per mitochondria. Therefore, the variation of mtDNA genomes among humans is relatively small and the power of discrimination is limited compared to the analysis of autosomal chromosomes. The mtDNA genome is approximately 16 569 bp long and it was sequenced already in 1981. It encodes 13 genes, two rRNAs and 22 tRNAs and has a 1100 bp non-coding sequence, known as the mtDNA control region, where the mutation rate is unusually high. Within the control region there are at least three hypervariable (HV) regions with a high number of SNPs (Brandstätter *et al.*, 2004) and these regions can be amplified by PCR and sequenced. In 1999, EDNAP's mitochondrial DNA population database project (EMPOP) (*www.empop.org*) was initiated with the purpose of creating a common forensic standard for mtDNA sequencing and an on-line mtDNA database with high-quality mtDNA population data. The initiative was a reaction to the significant number of errors in public mtDNA databases (Brandstätter *et al.*, 2004) and only forensic laboratories qualified by successful participation in EMPOP collaborative exercises have permission to submit mtDNA sequences (Parson *et al.*, 2004). Today, the typical targets for forensic and anthropological investigations are HV-I (position 16 024–16 400) and HV-II (position 44–340). However, there is a growing interest for SNPs in the mtDNA coding region, because certain sequences in HV-I and HV-II are very common (e.g. 7% of all Caucasians have the same HV1/HV2 haplotype). Several SNP panels have been proposed (Grignani *et al.*, 2006; Brandstätter *et al.*, 2006) and a selection of coding region SNPs for forensic caseworks are likely to be recommended in the near future.

6.6   Forensic DNA phenotyping

When the police initiate a search for a person based on a witness report, they usually put out a description of the person based on just a few facts, such as sex, age, height, weight, distinct features and the colours of hair, eyes and skin. Some of these traits (phenotypes) are genetically determined and can, in theory, be predicted by typing the person's DNA, if the genetic markers causing the phenotype are known. In the reverse situation, where the police have no witnesses and no suspect, but only a biological sample left at the crime scene, information on phenotypes can be pivotal for the police inquiry.

Special attention has been addressed to genetic markers determining the colours of hair, eyes and skin, mainly because animal models have been studied for decades and many genes involved in the regulation of melanin synthesis have been identified. Melanin is a complex polymer synthesized from the amino acid tyrosine via a number of toxic intermediates (Rees, 2003; Nordlund et al., 2006). The synthesis is performed in specialized compartments (organelles known as melanosomes) in the melanocytes to protect the cell from the destructive intermediates. The mature melanosomes filled with melanin are transported from the melanocytes to keratinocytes in the skin and hair, where melanin shields the body from damaging UV light. The regulatory mechanisms involved in melanocyte differentiation, melanin synthesis and melanosome transport are very complicated and not fully understood, and since genetic variations in any of the dozens of regulatory and functional proteins involved in these processes may influence melanin synthesis, there are many candidate genes to investigate and many possible combinations of alleles that eventually make up the haplotype responsible for a certain colour. The melanocortin 1 receptor (MC1R) is one of the key regulators of melanogenesis in humans and several SNPs in MC1R are associated with red hair colour, number of freckles, nevus count and increased risk of skin cancer (melanoma) (Sturm, 2002; Duffy et al., 2004). In one model, various combinations of seven SNPs in MC1R were shown to account for 67% of the red-haired individuals in an Australian population (Duffy et al., 2004) and a similar model was proposed for north Europeans (Flanagan et al., 2000). Four SNPs in two of the genes commonly deleted in albinos, OCA2 and MATP (two membrane proteins without any known function), are associated with dark colours of hair, skin and eyes in Caucasians (Duffy et al., 2004; Graf et al., 2005), and recently an SNP in the SLC24A5 gene (encoding a putative Na^+/Ca^+ exchanger) was estimated to account for 30% of the difference in skin melanin between Caucasians and Africans (Lamason et al., 2005). Many of these SNPs have large differences in allele frequencies in different populations and some of them were originally categorized as ancestry informative markers (AIMs), which can be used to divide the entire human population into groups of ethnical origin. This is not particularly surprising, since one of the most prominent traits distinguishing major population groups is the colour of the skin, and it has been suggested to use AIMs for the prediction of traits related to ethnic origin instead of identifying and typing the causative markers (Frudakis et al., 2003a). However, for forensic purposes, it will be more prudent to report on genetic markers with known effects on specific phenotypes. Today, there is insufficient knowledge about such markers, but after the completion of the human genome (sequencing) project, genetic markers for complex human traits are rapidly being identified and it is expected that tools for the prediction of human phenotypes can become sufficiently accurate to be used in forensic investigations.

6.7   Ethical considerations of SNP genotyping

The information deduced from a DNA profile has been an important issue of discussion in the last two decades, and the fact that the CODIS STR loci do not reveal anything about race, intelligence, physical characteristics or possible genetic disorders of an individual has been important for the appliance to ethical criteria imposed on forensic DNA investigations in some countries (Koops and Schellekens, 2006). With the introduction of SNP typing in forensic genetics, these criteria may be challenged, because SNPs can reveal information that may be considered to be sensitive or private. For example, the ethnic origin of a person can be determined with a high degree of certainty by analysing AIMs or Y chromosome SNPs (Frudakis *et al.*, 2003b; Jobling and Tyler-Smith, 2003; Phillips *et al.*, 2006). This information could be very important for the police investigation, but for some people it may be problematic if the DNA profile reveals that a person has different ancestors than the person presumed, and the public may express concern that 'ethnic SNPs' are used to criticize a certain population in the community. On the other hand, if the typing of AIMs is prohibited by law, it will not be possible to type for certain human phenotypes (e.g. skin colour), because such traits are used to distinguish between populations and an important tool for law enforcement may be lost.

The advantages of using SNP markers for routine caseworks are so large that there is no doubt that SNP typing will be employed by forensic laboratories in the future. Whether typing of SNPs in coding regions and forensic DNA phenotyping should be permitted is an issue that needs to be addressed by the public and the forensic genetic community, and most likely it has to be discussed continuously in the years to come as more and more genetic markers are discovered and associated with distinct phenotypes.

6.8   References

Asamura, H., Uchida, R., Takayanagi, K., Ota, M. and Fukushima, H. (2006) Allele frequencies of six miniSTR loci in a population from Japan. *Int. J. Legal Med.* 120: 182–184.

Brandstätter, A., Peterson, C.T., Irwin, J.A., Mpoke, S., Koech, D.K., Parson, W. and Parsons, T.J. (2004) Mitochondrial DNA control region sequences from Nairobi (Kenya): Inferring phylogenetic parameters for the establishment of a forensic database. *Int. J. Legal Med.* 118: 294–306.

Brandstätter, A., Salas, A., Niederstätter, H., Gassner, C., Carrecedo, A. and Parson, W. (2006) Dissection of mitochondrial superhaplogroup H using coding region SNPs. *Electrophoresis* 27: 2451–2460.

Budowle, B., Allard, M.W., Wilson, M.R. and Chakraborty, R. (2003) Forensics and mitochondrial DNA: Applications, debates, and foundations. *Annu. Rev. Genom. Hum. Genet.* 4: 119–141.

Budowle, B., Moretti, T.R., Baumstark, A.L., Defenbaugh, D.A. and Keys, K.M. (1999) Population data on the thirteen CODIS core short tandem repeat loci in African

Americans, U.S. Caucasians, Hispanics, Bahamians, Jamaicans, and Trinidadians. *J. Forensic Sci.* **44**: 1277–1286.

Coble, M.D. and Butler, J.M. (2005) Characterization of new miniSTR loci to aid analysis of degraded DNA. *J. Forensic Sci.* **50**: 43–53.

Dixon, L.A., Murray, C.M., Archer, E.J., Dobbins, A.E., Koumi, P. and Gill, P. (2005) Validation of a 21-locus autosomal SNP multiplex for forensic identification purposes. *Forensic Sci. Int.* **154**: 62–77.

Duffy, D.L., Box, N.F., Chen, W., Palmer, J.S., Montgomery, G.W., James, M.R., Hayward, N.K., Martin, N.G. and Sturm, R.A. (2004) Interactive effects of MC1R and OCA2 on melanoma risk phenotypes. *Hum. Mol. Genet.* **13**: 447–461.

Flanagan, N., Healy, E., Ray, A., Phillips, S., Todd, C., Jackson, I.J., Birch-Machin, M.A. and Rees, J.L. (2000) Pleiotropic effects of the melanocortin 1 receptor (MC1R) gene on human pigmentation. *Hum. Mol. Genet.* **9**: 2531–2537.

Frudakis, T., Thomas, M., Gaskin, Z., Venkateswarlu, K., Chandra, K.S., Ginjupalli, S., Gunturi, S., Natrajan, S., Ponnuswamy, K. and Ponnuwamy, K.N. (2003a) Sequences associated with human iris pigmentation. *Genetics* **165**: 2071–2083.

Frudakis, T., Venkateswarlu, K., Thomas, M.J., Gaskin, Z., Ginjupalli, S., Gunturi, S., Ponnuswamy, V., Natarajan, S. and Nachimuthu, P.K. (2003b) A classifier for the SNP-based inference of ancestry. *J. Forensic Sci.* **48**: 771–778.

Gill, P. (2001) An assessment of the utility of single nucleotide polymorphisms (SNPs) for forensic purposes. *Int. J. Legal Med.* **114**: 204–210.

Gill, P., Brenner, C.H., Buckleton, J.S., Carrecedo, A., Krawczak, M., Mayr, W.R., Morling, N., Prinz, M., Schneider, P.M. and Weir, B.S. (2006) DNA commission of the International Society of Forensic Genetics: Recommendations on the interpretation of mixtures. *Forensic Sci. Int.* **160**: 90–101.

Gill, P., Jeffreys, A.J. and Werrett, D.J. (1985) Forensic application of DNA 'fingerprints'. *Nature* **318**: 577–579.

Graf, J., Hodgson, R. and van Daal, A. (2005) Single nucleotide polymorphisms in the MATP gene are associated with normal human pigmentation variation. *Hum. Mutat.* **25**: 278–284.

Grignani, P., Peloso, G., Achilli, A., *et al.* (2006) Subtyping mtDNA haplogroup H by SNaPshot minisequencing and its application in forensic individual identification. *Int. J. Legal Med.* **120**: 151–156.

Inagaki, S., Yamamoto, Y., Doi, Y., Takata, T., Ishikawa, T., Imabayashi, K., Yoshitome, K., Miyaishi, S. and Ishizu, H. (2004) A new 39-plex analysis method for SNPs including 15 blood group loci. *Forensic Sci. Int.* **144**: 45–57.

International HapMap Consortium (2005) A haplotype map of the human genome. *Nature* **437**: 1299–1320.

Jeffreys, A.J., Wilson, V. and Thein, S.L. (1985) Individual-specific 'fingerprints' of human DNA. *Nature* **316**: 76–79.

Jobling, M.A. and Tyler-Smith, C. (2003) The human Y chromosome: An evolutionary marker comes of age. *Nat. Rev. Genet.* **4**: 598–612.

Kidd, K.K., Pakstis, A.J., Speed, W.C., *et al.* (2006) Developing a SNP panel for forensic identification of individuals. *Forensic Sci. Int.* **164**: 20–32.

Koops, B. and Schellekens, M. (2006) *Forensic DNA phenotyping: Regulatory issues.* Submitted for publication.

Krawczak, M. (1999) Informativity assessment for biallelic single nucleotide polymorphisms. *Electrophoresis* **20**: 1676–1681.

Kwok, P. (2001) Methods for genotyping single nucleotide polymorphisms. *Annu. Rev. Genom. Hum. Genet.* **2**: 235–258.

Lai, E. (2006) Application of SNP technologies in medicine: Lessons learned and future challenges. *Genome Res.* **11**: 927–929.

Lamason, R.L., Mohideen, M.P.K., Mest, J.R., *et al.* (2005) SLC24A5, a putative cation exchanger, affects pigmentation in zebrafish and humans. *Science* **310**: 1782–1786.

Nordlund, J.J., Boissy, R.E., Hearing, V.J., King, R.A. and Ortonne, J.P. (2006) *The Pigmentary System: Physiology and Pathophysiology* (2nd edn), Oxford University Press, New York.

Opel, K.L., Chung, D.T., Drabek, J., Tatarek, N.E., Jantz, L.M. and McCord, B.R. (2006) The application of miniplex primer sets in the analysis of degraded DNA from human skeletal remains. *J. Forensic Sci.* **51**: 351–356.

Pakendorf, B. and Stoneking, M. (2005) Mitochondrial DNA and human evolution. *Annu. Rev. Genom. Hum. Genet.* **6**: 165–183.

Parson, W., Brandstätter, A., Alonso, A., *et al.* (2004) The EDNAP mitochondrial DNA population database (EMPOP) collaborative exercises: Organisation, results, and perspectives. *Forensic Sci. Int.* **139**: 215–226.

Phillips, C., Sanchez, J.J., Fondevila, M., *et al.* (2006) *Inferring geographic origin using autosomal ancestry informative markers in a single multiplex assay.* Submitted for publication.

Rees, J.L. (2003) Genetics of hair and skin colour. *Annu. Rev. Genet.* **37**: 67–90.

Reich, D.E., Schaffner, S.F., Dalby, M.J., McVean, G., Mullikin, J.C., Higgins, J.M., Richter, D.J., Lander, E.S. and Altshuler, D. (2002) Human genome sequence variation and the influence of gene history, mutation and recombination. *Nat. Genet.* **32**: 135–142.

Sachidanandam, R., Weissman, D., Schmidt, S.C., *et al.* (2001) A map of human genome sequence variation containing 1.42 million single nucleotide polymorphisms. *Nature* **409**: 928–933.

Sanchez, J.J., Børsting, C. and Morling, N. (2005) Typing of Y chromosome SNPs with multiplex PCR methods. In: *Forensic DNA Typing Protocols, Methods in Molecular Biology*, Vol. 297, Humana Press, Totowa, USA, pp. 209–228.

Sanchez, J.J., Phillips, C., Børsting, C., *et al.* (2006) A multiplex assay with 52 single nucleotide polymorphisms for human identification. *Electrophoresis* **27**: 1713–1724.

Sobrino, B., Brion, M. and Carracedo, A. (2005) SNPs in forensic genetics: A review on SNP typing technologies. *Forensic Sci. Int.* **154**: 181–194.

Sturm, R.A. (2002) Skin colour and skin cancer – MC1R, the genetic link. *Melanoma Res.* **12**: 405–416.

Wang, X., Tomso, D.J., Liu, X. and Bell, D.A. (2005) Single nucleotide polymorphism in transcriptional regulatory regions and expression of environmentally responsive genes. *Toxicol. Appl. Pharmacol.* **207**: S84–S90.

7

The X chromosome in forensic science: past, present and future

Reinhard Szibor

7.1   Introduction

The sex chromosomes or gonosomes, chromosome X (ChrX) and chromosome Y (ChrY) are unique and differ in several aspects from the other chromosomes, which are referred to as autosomes (AS). Both ChrX and ChrY are unique with regard to the major content of their genes and sequences. In the cells of normal human males not affected by chromosomal aberrations, sex chromosomes do not occur in pairs. Males carry one X and one Y chromosome. Hence, most ChrX and ChrY regions are hemizygous in males. However, blocks of sequence homology between X and Y chromosomes suggest a common origin. During male gametogenesis, recombination between X and Y chromosomes occurs in small sub-telomeric regions of the X and Y chromosomes called the pseudoautosomal regions. These segments are homologous. Genes and markers in the pseudoautosomal region are not sex-linked. Recombination frequencies in this region are 20 times higher than on autosomes.

There are two pseudoautosomal regions – the Xp and Xq telomeres – referred to as PAR1 and PAR2. Furthermore, ChrX and ChrY show several regions of homology in addition to the common pseudoautosomal regions. In females ChrX is present as a homologous pair and resembles autosomes in this respect. However, even individuals with more than one ChrX possess only one active ChrX per cell. According to the Lyon Hypothesis (Lyon, 1961) additional copies are inactivated, which explains why ChrX monosomies, trisomies and polysomies are compatible with life. Functional ChrX inactivation is connected with the formation of a morphologically visible heteropycnic chromatin that is also

Molecular Forensics.   Edited by Ralph Rapley and David Whitehouse
Copyright 2007 by John Wiley & Sons, Ltd.

called sex chromatin. Inactivated ChrXs are visible in many but not in all female cells as Barr bodies (Barr and Bertram, 1949; Barr and Carr, 1962) or 'drumsticks' (Davidson and Smith, 1954). The latter are structures in the nuclei of polymorphonuclear leucocytes. Chromosome X monosomies, trisomies and other polysomies occur in different forms of appearance and may be connected with serious handicaps and infertility. Triple X (trisomy X) females frequently have a nearly normal development.

For parental generations, such gonosomal irregularities can usually be excluded since they would be associated with infertility. Unexpected and undetected aberrant gonosomal karyotypes in an offspring may however occur and affect the accuracy of kinship testing using ChrX markers. Gonosomal genotype X0, for example, which is associated with Ullrich-Turner syndrome, occurs at an incidence of 1 in 2500 female live-births (Clement-Jones et al., 2000). Both complete and partial monosomies have been observed. Another unexpected situation is when an XY-karyotype occurs in phenotypic females (Wieacker et al., 1998). This happens in context with androgen insensitivity or XY gonadal dysgenesis. Such disturbances cause genetic males to present with an unobtrusive female phenotype, although the present of ChrY can easily be detected by an amelogenin test. The posterior probability of a full or partial ChrX monosomy, or an XY female, increases when several closely linked ChrX markers appear to be homozygous. In the way that they perturb kinship testing, karyotypes XO and female XY are formally equivalent to autosomal uniparental disomy (Bein et al., 1998; Wegener et al., 2006). As with AS markers, paternity exclusion that relies upon ChrX marker homozygosity thus requires independent experimental verification.

A gonosomal aberrant male karyotype with XXY (or XXXY, XXXXY, etc.) develops Klinefelter syndrome and shows a prevalence of about 1:500 males (Bojesen et al., 2003). Klinefelter syndrome may be detected in kinship testing when ChrX markers show heterozygosity. Ethical aspects are discussed below.

7.2 History of forensic utilization of the X chromosome

The fundamental idea for extensive usage of X-chromosomal markers in forensic practice came from the experiences made during the second half of the last century in the field of clinical genetics. There are many well-known diseases and traits such as haemophilia, Duchenne muscular dystrophy, Lesch-Nyhan syndrome, G6PD deficiency, colour blindness, etc. that follow X-chromosomal inheritance. If a male patient is fertile, all his daughters possess the defective paternal X chromosome and transmit it to half of the next generation. Half of all daughters are again gene carriers and also half of their sons inherit the defective allele and exhibit the trait due to the hemizygote state of their ChrX. Furthermore, when a male exhibits two or more ChrX-linked traits it is obvious that alleles of all relevant loci are unified to one haplotype. This explains why ChrX linkage analysis (Adam *et al.*, 1963) is fairly easy. It is obvious that knowledge of such simple contexts is valuable not in clinical genetics only but also in kinship testing. However, cognition of sex-linked genetic markers usable in forensic genetics rose very slowly.

In the course of forensic kinship testing, which started in the middle of the last century, initially only blood group markers played a role. These were later supplemented by serum protein and enzyme variants. However, all of them were of autosomal inheritance. Regarding X-chromosomal markers, the first significant achievement was made when the Xg^a blood group was detected by the team of Race and Sanger (Mann *et al.*, 1962; Sanger *et al.*, 1971, 1977; Tippett and Ellis, 1998; Went *et al.*, 1969). Chromosome X linkage of Xg^a could easily be recognized by comparing the frequencies of the Xg^a/Xg phenotypes in males and females (males: 0.62/0.38; females: 0.86/0.14). The blood group Xg^a is a fairly weak antigen. This may be the reason why Xg^a testing could not be established as a significant method in kinship testing practice. However, some questions of scientific interest, such as identification of the origin of X chromosomes in chromosome aberration syndromes such as Klinefelter and Ullrich-Turner syndrome have been solved using serological Xg^a testing (Mann *et al.*, 1962; Lewis *et al.*, 1964; Tippett and Ellis, 1998). Later Ellis *et al.* (1994) identified the Xg^a antigen derived from the N-terminal domain of a candidate gene, referred to earlier as PBDX.

Two further ChrX loci encoding gene products with polymorphic appearance and therefore with a potential for usage in kinship testing are known. Glucose-6-phosphate dehydrogenase (G6PD) (Adam *et al.*, 1967; Yoshida *et al.*, 1973; Askov *et al.*, 1985; Roychoudhury and Nei, 1988) and phosphoglycerate kinase (PGK) (Beutler, 1969; Chen *et al.*, 1971; Roychoudhury and Nei, 1988) show considerable diversity on the protein level in some geographical regions. However, to our knowledge, with rare exceptions (Bucher and Elston, 1975), X-linked enzyme variants do not play any role in forensic contexts.

In the pre- DNA-technique era forensic scientists used sex chromatin tests for gender assessment in human tissues or single cells (Given, 1976; Duma and Boskovski, 1977). Later this technique was complemented by fluorescence microscopic demonstration of male heterochromatin (Tröger *et al.*, 1976; Mudd, 1984). It can be shown as a quinacrine mustard-stained part of ChrY in metaphases and even in metaphase cell nuclei.

In the early 1980s clinical geneticists started with genomic linkage analysis aimed at gene carrier detection and prenatal diagnosis (Davies *et al.*, 1963; Davies *et al.*, 1985). Some X-linked diseases, such as Duchenne muscular dystrophy and haemophilia, were in the focus of interest. In the first stage typing targets mainly were single nucleotide polymorphisms (SNPs), which could be detected by the Southern technique and were called restriction fragment length polymorphisms (RFLPs) according to the detection technique involving restriction enzymes. Later SNP linkage markers were supplemented by CA repeat polymorphisms (Lalloz *et al.*, 1991, 1994) and the minisatellite St14 (DXS52) (Oberle *et al.*, 1985). Investigation of the latter kind of polymorphism was enabled by creating the polymerase chain reaction (PCR) technique. Whilst usage of dinucleotide repeats is shunned by the forensic community, minisatellite marker DXS52 clearly fulfills the forensic requirements for markers. Nevertheless, DXS52 appeared in the forensic literature only very sporadically (Lambropoulos *et al.*, 1995; Yun and Yun, 1996). The first two ChrX microsatellites that played a significant role were HPRTB (Hearne and Todd, 1991; Edwards *et al.*, 1992; Kishida *et al.*, 1997; Xiao *et al.*, 1998; Szibor *et al.*, 2000) and ARA (Edwards *et al.*, 1992; Sleddens *et al.*, 1992; Kishida and Tamaki, 1997; Desmarais *et al.*, 1998).

Kishida *et al.* (1997) and Desmarais *et al.* (1998) almost at the same time created formulas for the calculation of useful parameters such as mean exclusion chance (MEC), which considers the unique inheritance of ChrX. Thirty years before Krüger *et al.* (1968) had created the MEC for AS. These formulas are summarized in Table 7.1.

One of the challenges in kinship testing is to establish techniques that can bridge large pedigree gaps. We know from observation in clinical genetics that persons who share a very rare genetic feature can be unified to a common pedigree. The famous monarchic haemophilia pedigree can be demonstrated as an example. Some members of the European high nobility show the trait of haemophilia, which leads to the term 'royal disease'. Typing of the haemophilia gene would enables us to show that all affected persons with a certain mutation belong to the same pedigree descending from Queen Victoria (1837–1901). Two reasons ban us from doing this. Firstly, typing of harmful mutations in kinship

Table 7.1 Formulas for evaluation of the forensic efficiency of genetic markers

No.[a]	Formula[b]	References
I	$\sum_i f_i^3 (1-f_i)^2 + \sum_i f_i (1-f_i)^3 + \sum_{i<j} f_i f_j (f_i + f_j)(1-f_i-f_j)^2$	Krüger *et al.* (1968)
II	$\sum_i f_i^3 (1-f_i) + \sum_i f_i (1-f_i)^2 + \sum_{i<j} f_i f_j (f_i + f_j)(1-f_i-f_j)$	Kishida *et al.* (1997)
III	$1-\sum_i f_i^2 + \sum_i f_i^4 - \left(\sum_{i<j} f_i^2\right)^2$	Desmarais *et al.* (1998)
IV	$1-2\sum_i f_i^2 + \sum_i f_i^3$	Desmarais *et al.* (1998)
V	$1-2\left(\sum_i f_i^2\right)^2 + \sum_i f_i^4$	Desmarais *et al.* (1998)
VI	$1-\sum_i f_i^2$	Desmarais *et al.* (1998)

[a] I: MEC (mean exclusion chance) for AS markers in trios; II: MEC for ChrX markers in trios involving daughters; III: MEC for ChrX markers in trios involving daughters; IV: MEC for ChrX markers in father/daughter duos; V: power of discrimination (PD) in females; VI: PD for ChrX markers in males.
[b] $f_i(f_j)$: population frequency of the *i*th (*j*th) marker allele.

testing would violate our ethical principles. Secondly, very rare traits would contribute to kinship only very seldomly. However, this example can demonstrate the power of rare alleles. Fortunately, the same effect can be achieved in another way: substituting short tandem repeats (STRs) by haplotypes consisting of clustered STRs provides a comparable power and can be used systematically in kinship testing.

7.3 Chromosome X short tandem repeats

Microsatellites or simple sequence repeats are tandemly repeated DNA sequences found in varying abundance in all human chromosomes. In forensic science the term short tandem repeat (STR) is ingrained. Including ChrX the overall STR density is comparable in all chromosomes (Subramanian *et al.*, 2003). Within the chromosomes the density of STRs, however, shows significant variations. Tri- and hexanucleotide repeats are more abundant in exons, whereas other repeats are more abundant in non-coding regions. Moreover, as has been shown (McNeil *et al.*, 2006), a striking enrichment (>10-fold) of $[GATA]_n$ is revealed throughout a 10 Mb segment at Xp22 that escapes inactivation (Lyon, 1961), and is confirmed by fluorescence *in situ* hybridization. A similar enrichment is found in other eutherian genomes. These findings clearly demonstrate sequence differences relevant to the novel biology and evolution of the X chromosome. Furthermore, they implicate simple sequence repeats, linked to gene regulation and unusual DNA structures, in the regulation and formation of facultative heterochromatin.

The analysis of tri-, tetra- and penta-STRs has become widespread in forensic medicine and STRs located on autosomes (Brinkmann, 1998; Urquhart *et al.*, 1994) were used long before application of Y-chromosomal (de Knijff *et al.*, 1997; Jobling *et al.*, 1997; Kayser *et al.*, 1997a, 1997b; Roewer *et al.*, 2001) and X-chromosomal STR markers. Although the existence of ChrX STRs, i.e. HPRTB (Hearne and Todd, 1991; Edwards *et al.*, 1992) and ARA (Edwards *et al.*, 1992; Sleddens *et al.*, 1992) and DXS981 (Mahtani and Willard, 1993), was reported relatively early, the desire to use such markers as tools for forensic application came up later.

The International Society for Forensic Haemogenetics guidelines for the forensic use of microsatellite markers (Bär *et al.*, 1997) apply to both AS and ChrX STRs. However, some specific molecular and formal genetic aspects need to be taken into account when dealing with ChrX markers. The forensic application of microsatellite markers can be done in practise if they are in Hardy–Weinberg equilibrium and have a high enough degree of polymorphism. The application of coding STRs such as ARA should be avoided. Some ethical considerations are discussed below.

Figure 7.1 and Table 7.2 review the main forensic ChrX repeat markers known to date. Most of them show no specific peculiarities in terms of their

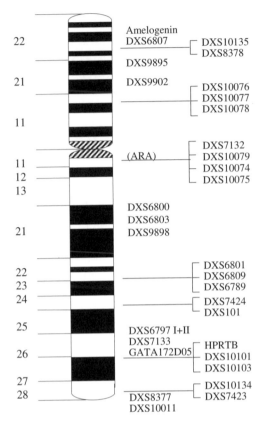

Figure 7.1   Chromosome X ideogram with STR markers

handling and some are routinely used by our own group and others. Solely DXS10011 should be singled out. This tetranucleotide repeat marker occurs in two sequence variants, type A and B. Whereas the type B allele mutation rate is comparable with other STRs, type A alleles mutate with very high frequencies (Hering *et al.*, 2004). In general, mutation rates of ChrX STRs seem not to differ from microsatellites of other chromosomes. An early report on a high rate of mutation in DXS981 (Mahtani and Willard, 1993) could not be confirmed in kinship testing and seems to be a misinterpretation of results obtained when working with lymphoblastoid cell lines of CEPH families (Banchs *et al.*, 1994).

Typical STRs show regular structure with constant length differences between different alleles, i.e. 3, 4 or 5 bp steps. However, several markers such as DXS7130 and DXS6803 (Edelmann and Szibor, 2003), DXS10011 (Watanabe *et al.*, 2000; Hering *et al.*, 2004) and DXS981 (Mahtani and Willard, 1993; Tabbada *et al.*, 2005) are composed of different repeat units resulting in 2 or

Table 7.2 Characteristics of ChrX STRs in forensic use based on allele distribution in a European population

STR[a]	Physical and genetic localization[b] (distance in bp and cM from Xp telanere)		Het[c]	MEC(II)[d] (trio case)	MEC(IV)[d] (duo case)	PD (female)[d]	PD (male)[d]	Reference
DXS6807 (Tetra)	4 603 118	4.39	0.608	0.709	0.471	0.838	0.671	Edelmann and Szibor (1999)
DXS9895 (Tetra)	7 236 843	8.76	0.694	0.704	0.554	0.886	0.741	Edelmann et al. (2001)
DXS10135 (Tetra)	9 116 057	No report	0.9251	0.9159	0.8499	0.9883	0.9214	Szibor, unpublished
DXS8378 (Tetra)	9 179 962	No report	0.658	0.714	0.532	0.868	0.719	Edelmann et al. (2001)
DXS9902 (Tetra)	15 083 273	22.04	0.636	0.743	0.490	0.848	0.695	Edelmann et al. (2001)
DXS10076 (Tetra)	48 065 558	No report	0.784	0.751	0.620	0.921	0.782	Augustin et al. (2006)
DXS10077 (Tri)	48 073 264	No report	0.518	0.468	0.322	0.717	0.516	Augustin et al. (2006)
DXS10078 (Tetra)	48 078 442	No report	0.825	0.799	0.681	0.945	0.823	Augustin et al. (2006)
DXS7132 (Tetra)	64 438 357	83.30	0.687	0.883	0.557	0.883	0.731	Edelmann et al. (2001)
DXS10074 (Tetra)	66 760 138	No report	0.833	0.811	0.7059	0.951	0.832	Hering et al. (2006)
DXS981 (Tetra)	67 980 377	No report	0.8426	0.8171	0.7055	0.9532	0.8374	Tabbada et al. (2005)
DXS6800 (Tetra)	7 486 555	93.17	0.690	0.694	0.548	0.868	0.729	Edelmann et al. (2001)
DXS9898 (Tetra)	87 602 564	No report	0.731	0.745	0.596	0.908	0.769	Hering and Szibor (2000)
DXS6801 (Tetra)	92 317 317	99.73	0.6472	0.582	0.4535	0.816	0.627	Edelmann and Szibor (2005)
DXS6809 (Tetra)	94 744 298	102.28	0.808	0.815	0.7033	0.953	0.835	Edelmann et al. (2003)
DXS6789 (Tetra)	95 255 559	103.56	0.746	0.702	0.5677	0.893	0.741	Hering et al. (2001)

Table 7.2 Continued

STR[a]	Physical and genetic localization[b] (distance in bp and cM from Xp telanere)		Het[c]	MEC(II)[d] (trio case)	MEC(IV)[d] (duo case)	PD (female)[d]	PD (male)[d]	Reference
DXS7424 (Tri)	100 424 961	No report	0.764	0.836	0.639	0.928	0.794	Edelmann et al. (2002)
DXS101 (Tri)	101 219 161	No report	0.78	0.885	0.794	0.978	0.889	Edelmann and Szibor (2001)
DXS6797 (Tetra)	107 287 210	112.89	0.754	0.712	0.575	0.898	0.753	Poetsch et al. (2005b)
DXS7133 (Tetra)	108 847 688	No report	0.575	0.658	0.422	0.800	0.635	Edelmann et al. (2001)
GATA172D05 (Tetra)	112 980 738	116.17	0.775	0.804	0.654	0.935	0.808	Edelmann et al. (2001)
HPRTB (Tetra)	133 341 004	No report	0.737	0.919	0.610	0.919	0.779	Poetsch et al. (2005a)
DXS9908 (Tetra)	142 666 846	165.11	0.760	0.720	0.586	0.906	0.754	Edelmann et al. (2006)
DXS8377 (Tetra)	149 237 039–	No report	0.916	0.922	0.855	0.989	0.924	Edelmann et al. (2001)
DXS10134 (Tetra)	149 320 642	No report	0.859	0.842	0.741	0.964	0.857	Edelmann, unpublished
DXS7423 (Tetra)	149 381 471	No report	0.688	0.734	0.548	0.884	0.734	Szibor et al. (2003)
DXS10011 (Tetra)	150 858 594	No report	0.964	0.944	0.827	0.995	0.947	Hering et al. (2004)

[a] 'Tri' and 'Tetra' refer to the repeat type.
[b] Physical mapping data were retrieved from Human Genome Browser (http://www.genome.ucsc.edu/cgi-bin/hgGateway) and genetic localizations were obtained from the Marshfield database (http://www.marshfieldclinic.org/).
[c] Expected heterozygosity was calculated from Nei and Roychoudhury (1974).
[d] Definitions can be found in Table 7.1.

1 bp differences between alleles. Consequently, typing of such markers essentially requires gene scanner equipment and advanced carefulness.

Table 7.2 also contains information on the power of forensic analysis using the respective markers. Several of the population genetic parameters of forensic interest are shown for each STR listed, namely the expected heterozygosity (Nei and Roychoudhury, 1974), the MEC in trio and duo cases and the power of discrimination. The data shown here refer to a population sample of central Europe (Germany).

7.4 Power of ChrX markers in trace analysis

The power of discrimination (PD) value of ChrX markers varies according to gender. When female traces are to be matched to female individuals the ChrX markers equal those for AS (Table 7.1, formula V). For the matching of male traces to male suspects, the PD value of ChrX markers (Table 7.1, formula VI) is generally smaller than that of AS markers. This is due to the fact that male ChrX analysis utilizes only one allele per STR.

In a mixed female/male stain, the chance of having all male alleles included in the female component is higher for ChrX than for AS markers, therefore it is not advisable to use ChrX markers to test male traces in a female background. In order to identify female traces in male contamination, however, ChrX markers are more efficient than AS markers because the female alleles can only be completely included in the male component if the female coincidentally happens to be homozygous at all loci.

We conclude that, with few exceptions, ChrX markers are less powerful in stain analyses than AS markers. Furthermore, if more than four ChrX STRs are used in casework it is unavoidable that two or more of them are linked. Hence, due to the possibility of the existence of a linkage disequilibrium (LD), calculation of identity likelihood can be difficult. Nevertheless, demonstration of female skin debris under male finger nails or vaginal cells at a penis, etc. may be an issue for application of an X-chromosomal STR amplification kit.

7.5 Power of ChrX markers in kinship testing

Formulas developed specifically for ChrX markers in the context of kinship testing are listed in Table 7.1. They can be used to calculate the gain of the concerning marker in usage for kinship testing in dependence of the allele distribution in a certain population. In trios involving a daughter, ChrX markers are more efficient than AS markers. This fact is reflected by the formulas for calculation of MEC(I) for AS markers (Krüger *et al.*, 1968) and MEC(II) (Kishida *et al.*, 1997) and MEC(III) (Desmarais *et al.*, 1998). MEC(III) is equivalent to MEC(II). MEC(I) for AS is not suitable for ChrX markers except for deficiency cases in which the paternal grandmother is investigated instead of the alleged father. If MEC(I) is compared to MEC(II) and MEC(III), the latter are consistently larger. Finally, Desmarais *et al.* (1998) introduced formulas for the mean exclusion chance of ChrX markers involving father/daughter duos lacking maternal genotype information, MEC(IV), which is also appropriate for maternity testing of mother/son duos.

Paternity testing in trios and duos

Paternity cases involving the common trio constellation of mother, offspring and alleged father can usually be solved with AS STRs alone, and do not seem to require any additional or alternative markers. When father/son relationships are to be tested, ChrX markers can contribute nothing anyway. However, when father/daughter relationships are in question it may be worthwhile including ChrX markers in testing. This is especially the case when difficult-to-analyse template materials are involved, such as DNA from exhumed skeletons, historical or prehistorical samples, etc. Despite primer sets for typing degraded DNA now being available (Hellmann *et al.*, 2001; Wiegand and Kleiber, 2001; Asamura *et al.*, 2006; Meissner *et al.*, 2006), in such instances sufficient statistical power has to come from a small number of low-size STRs. Fortunately, ChrX STRs are normally characterized by relatively high MECs, even at a low to medium degree of polymorphism (Table 7.3). In those contexts, ChrX markers may be superior to AS markers in some instances. As an example I would like to mention a case that we have solved recently. We were requested to prove a father /daughter relationship by typing only the daughter's and the alleged father's saliva. The paternal saliva trace was taken from a stamp licked 30 years ago. The paternity likelihood could be established to be 99.93 by DXS8378-DXS7132, HPRTB and DXS7423 alone. Autosomal systems had contributed only little in solving this question.

Table 7.3   Comparison of the mean exclusion chance (MEC) of short amplicon STRs located at ChrX and autosome (AS) markers. For definition MEC (I)and MEC (II) see Table 1

ChrX marker	Mini STR product size	Het[a]	MEC(I)[b]	MEC(II)[c]	AS marker	Mini STR product size	Het[a]	MEC(I)[b]
DXS8378	110–134	0.733	0.497	0.687	D3S1358	121–141	0.80	0.595
DXS7132	131–155bp	0.737	0.497	0.688	TH01	66–92bp	0.776	0.5571
GATA172D05	108–136bp	0.8053	0.617	0.775	VWA	134–162	0.810	0.6229
Cumulative	–	–	**0.905**	**0.976**			–	**0.935**

[a] Heterozygosity.
[b] See Table 7.1 for definition.

Paternity cases involving blood-relatives

In paternity cases involving close blood-relatives as suspects, the exclusion power of STRs is substantially reduced and ChrX STRs may be superior to AS markers. For example, if two alleged fathers are father and son, they would not share any X-chromosomal alleles identical by descent (ibd) so that ChrX markers would be more efficient than AS markers. Brothers, in contrast, share a given maternal ChrX allele with probability 0.5, which corresponds to the probability of exactly one allele shared identical by descent at an AS locus.

For four unlinked ChrX loci, the chance of sharing alleles identical by descent would be $0.5^4 = 0.0625$. Unfortunately, since the ChrX length is not more than 198 cM this chromosome can host a maximum of four, but strictly only three unlinked marker regions. When the markers are closely linked, they do not segregate independently. As with AS markers, they would instead represent a single haplotype that is shared with a probability of 0.5. The ChrX contains four linkage groups located at Xp22.2, Xq12, Xq26 and Xq28 that can provide independent genotype information (Figure 7.1). At present, we propose that it is preferable to use clusters DXS10135-DXS837, DXS7132- DXS10074, HPRTB-DXS10101 and DXS7423- DXS10134 to define haplotypes in forensic practise. Typing of these four marker pairs can be done by using the PCR kit Mentype® Argus X-8 that is now commercially available. Alternatively, other cluster haplotypes such as DXS101-77424, DXS6801-DXS6809-DXS6789 and DXS10076-DXS10077-DXS10078 may be chosen on the basis of Table 7.2 and Figure 7.1.

Paternity testing using abortion material in incest and rape cases

After incest or criminal sexual assault, pregnancies may be terminated by suction abortion. An aborted 6–8 week product of conception consists of small amounts of non-identifiable foetal organs as well as maternal blood and other tissues. In such cases, microscopic detection of embryonic organs or chorionic villi is not generally successful and samples will contain a mixture of foetal and maternal DNA. By typing the simple amelogenin dimorphism, the appearance of a ChrY signal can clarify the sex. For male foetuses, further ChrY testing can easily help to assess the paternity. In incest cases in which a father is charged with abusing his daughter, however, ChrX testing of the abortus cannot demonstrate paternity since all foetal alleles would necessarily coincide with alleles of the daughter. In such cases, ChrX testing of mixed abortion material can only be used for the purpose of exclusion, not inclusion. The highest certainty may be provided by a typing strategy using AS and ChrX markers simultaneously.

A quite different situation occurs when incest has to be investigated and clean foetal material can be obtained by chorion biopsy. In the case of a female foetus, ChrX testing would prove the fathering by a father–daughter incest when all foetal alleles coincide with alleles of the pregnant woman. Recently, we have reported a case of prenatal exclusion without involving the putative father of an incestuous father–daughter parenthood (Schmidtke *et al.*, 2004).

Maternity testing

There are some situations in which mother/child testing is requested. For example, public authorities responsible for aliens often allow family reunion only after proved kinship. Maternity can also be demonstrated by sequencing mitochondrial DNA (mtDNA), however this technique would not always yield the same level of certainty. For example, mtDNA sequences are identical not only to those of their own children but also to nephews and nieces in a maternal line. Furthermore, due to the high rate of illegitimate paternity in modern societies, the identification of skeletons or carcasses by mother/child testing is more reliable than through the assessment of father/child relationships.

Typing of ChrX STRs may thus represent a sensible alternative option to assess maternity. For testing mother–daughter relationships, ChrX markers are equivalent to AS markers and do not provide any specific advantage. Testing mother–son kinship, however, is more efficiently performed using ChrX markers. As discussed above, the option for ChrX marker typing using short amplicons (Asamura *et al.*, 2006) should be considered, especially when skeletal human remains or other difficult samples have to be analysed. The exclusion chance in such cases is identical to that of ChrX STRs in father/daughter tests (see Table 7.1).

7.6   Chromosome X marker mapping and haplotype analysis

Genetic investigations in paternity trio cases should involve at least 12 STRs located on 10 chromosomes. However, solving complex kinship cases needs additional effort and ChrY and ChrX haplotyping is proving to be a powerful tool in solving difficult questions. The simultaneous analysis of STRs located on the same chromosome requires knowledge about the extent of pairwise linkage and linkage disequilibrium between them. Comprehensive studies on ChrX sequencing have determined 99.3% of the euchromatic sequence of the X chromosome (Ross *et al.*, 2005). Hence, for a short time physical mapping of most ChrX markers, given as distance from the Xp telomere (measured in base pairs, bp), can be performed using internet databases such as the UCSC Genome Browser Database (Karolchik *et al.*, 2003). Unfortunately, however, genetic and physical distances are not strictly correlated (Nagaraja *et al.*, 1998).

The classical approach to studying linkage between markers is via pedigree analysis. Based upon LOD ('logarithm of the odds') scores calculated from family data (Ott, 1991), meiotic recombination fractions are estimated for pairs of markers and transformed into genetic distances (measured in centimorgan, cM) using appropriate mapping functions. Due to the hemizygosity of sex chromosomes in males, linkage analysis is particularly efficient for ChrX loci. Nevertheless, for accurate mapping of very short distances, typing of a high number of meioses would be necessary. On the other hand, due to the limitations in typing very high numbers of meioses it may be justified to employ a simple rule of thumb: a physical distance of 1 Mb corresponds to a genetic distance of 1 cM, i.e. one expected recombination per 100 meioses. Genetic data resources available via the Marshfield and NCBI websites can be consulted to access genetic localization for many markers. Figure 7.1 illustrates the distribution of some practically relevant ChrX STRs along the ChrX ideogram. Under practical aspects we have subdivided the ChrX into separate linkage groups 1–4. Under practical aspects, four STRs that are unlinked, i.e. DXS8378, DXS7132, HPRTB and DXS7423, have been chosen and can be seen as the cores of these four linkage groups. This set of four unlinked STRs (plus amelogenin) can be typed using the commercially available forensic ChrX typing kit (Mentype® Argus X-UL). In a second stage the kit is extended by further STRs. Mentype® Argus X-8 provides a valuable tool for ChrX haplotyping. Each of the four STR clusters spans less than 0.5 cM and therefore provides a stable haplotype. Consequently, the genetic risk of recombination within each of the four clusters will be less than 0.5%.

Generally, alleles of linked loci form haplotypes that recombine during meioses at a frequency corresponding to the inter-marker genetic distance. In kinship testing, haplotypes of closely linked STRs must therefore be analysed as a whole, rather than through their constituent alleles, if the meiotic stability of haplotypes is high enough. Linkage disequilibrium (LD), which refers to this 'non-random' association of alleles at different loci, measures the deviation of population-specific haplotype frequencies from the product of the corresponding allele frequencies. For markers with strong LD, haplotype frequencies cannot be inferred from allele frequencies alone but instead have to be estimated directly from population data. Due to their higher mutation rate, STRs tend to show less LD than SNPs. However, LD can still occur between closely linked STRs and therefore has to be assessed prior to their practical application. In some ChrX regions we analysed the inter-marker LD of STRs by genotyping male DNA samples. Significant LD was observed for some very tight-linked marker clusters, namely DXS101 and DXS7424 (Edelmann et al., 2002), DXS6801, DXS6809 and DXS6789 (Szibor et al., 2005b), DXS10076, DXS10077 and DXS10078 (Augustin et al., 2006) and DXS10079, DXS10074 and DXS10075 (Hering et al., 2006). For example, testing the latter mentioned cluster in an Eastern German population of 781 unrelated men revealed 172 different haplotypes. Due to a considerable LD the number of observed haplotypes is smaller than previously expected. Nevertheless, 72% of all observed haplotypes showed frequencies of <0.02. Hence, despite the existence of LD, these STRs and the other

STR clusters mentioned are characterized by a high evidentiary power in kinship testing.

Deficiency paternity cases

When, in a case of uncertain paternity, a biological sample from a putative father is not available and DNA from paternal relatives has to be analysed instead, the situation is called a 'deficiency paternity test'. This branch of kinship testing is the main field of ChrX marker application and the major advantage of ChrX markers can be demonstrated here. Again, as illustrated using the royal pedigree, X-chromosomal traits or ChrX marker testing can connect pedigree members through large distances with respect to X-chromosomal tracks, however they fail when X-chromosomal lines are interrupted by father–son relationship. The benefit of ChrX testing can be shown by the presentation of some examples.

When ChrX markers are investigated in a deficiency case, the mother of the unavailable putative father (i.e. the putative grandmother) is the key figure. Instances in which she is available for genotyping, strictly speaking, do not represent deficiency cases. All ChrX alleles of the putative father can be determined by investigating her, and the MEC can be calculated using the respective formula for AS markers (Krüger et al., 1968; Table 7.1). The ChrX marker genotype of the putative grandmother can also be reconstructed to some extent from her children. If she has several daughters, it is possible to determine the parental origin of most of their ChrX alleles and therefore the grandmaternal genotype. If brothers of the putative father are available, the data are even more informative (Figure 7.2). Then, the grandmaternal genotype must have been heterozygous for all ChrX loci for which brothers of the putative father carry different alleles. If they carry identical alleles, the constellation is uninformative: the mother can be either homozygous or heterozygous at the corresponding locus. If closely linked loci have already been identified as being heterozygous, the probability of homozygosity at the original locus can be assessed by haplotyping. This is exemplified in Figure 7.2 for DXS6801, DXs6809 and DXS6789. In this case a woman 'Nora', and her putative uncles 'Jim' and 'Joe' were tested for kinship. We typed 18 ChrX markers, including the Xp21 cluster (Szibor et al., 2005b). For some loci the constellation is informative in the sense that it narrows down the set of possible alleles of the putative grandmother (PGM) and consequently of the putative father (PF). In the present case, single STR typing revealed such a pattern and a consequent exclusion of paternity for DXS8378 and DXS10011. However, since DXS10011 is strongly prone to mutation, the exclusion was still regarded as weak. Anyhow, whilst single STR results for the Xp21 cluster were uninformative, haplotyping showed that the necessary paternal allele combination of 14–31–14 could not have been inherited from the PF. For the PF to have received this haplotype from the PGM,

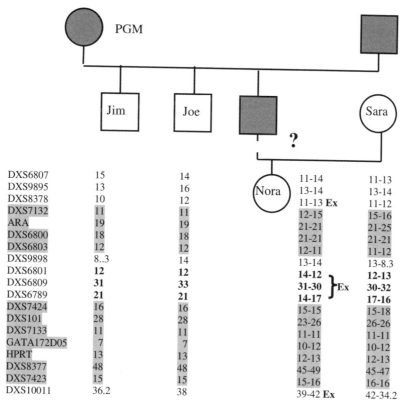

DXS6807	15	14	11-14	11-13
DXS9895	13	16	13-14	13-14
DXS8378	10	12	11-13 **Ex**	11-12
DXS7132	11	11	12-15	15-16
ARA	19	19	21-21	21-25
DXS6800	18	18	21-21	21-21
DXS6803	12	12	12-11	11-12
DXS9898	8..3	14	13-14	13-8.3
DXS6801	12	12	**14-12**	**12-13**
DXS6809	31	33	**31-30** }**Ex**	**30-32**
DXS6789	21	21	**14-17**	**17-16**
DXS7424	16	16	15-15	15-18
DXS101	28	28	23-26	26-26
DXS7133	11	11	11-11	11-11
GATA172D05	7	7	10-12	10-12
HPRT	13	13	12-13	12-13
DXS8377	48	48	45-49	45-47
DXS7423	15	15	15-16	16-16
DXS10011	36.2	38	39-42 **Ex**	42-34.2

Figure 7.2 Deficiency paternity test. The ChrX typing at loci that are not shaded in grey is fully informative and reveals the grandmaternal genotype. Since the likelihood for a double crossing-over within the DXS6801-DXS6809-DXS6789 cluster is very low, haplotyping provides the third X-chromosomal paternity exclusion for the putative father

two recombinations would have been required. Since the probability of such a double recombination is only $0.051 \times 0.046 = 2.35 \times 10-3$, i.e. of the same order of magnitude as the mutation rate of most STRs, the PF could unequivocally be excluded from paternity.

When female individuals have the same father, they also share the same paternal ChrX. An investigation of ChrX markers of two sisters or step-sisters can thus exclude paternity even when none of the parents are available for testing. The AS markers cannot provide such information. A positive proof of paternity is also possible with a lack of maternal genotype information, but is generally less reliable. This is due to the fact that sisters usually inherit only partially matching haplotypes from their mother. The co-inheritance of two identical maternal ChrXs without a recombination is not impossible, but rare. With a total genetic length of approximately 198 cM (Deloukas et al., 1998) there are several virtually uncoupled regions on the ChrX (Figure 7.1). Assuming that the number of recombination breakpoints between two ChrX loci follows a Poisson distribution with parameter λ equal to the genetic distance between the loci (Haldane, 1919), the chance of inheriting a non-recombined ChrX equals $e^{-2} = 0.135$ ($200\,\text{cM} = 2\,\text{M}$, the basic unit of genetic distance). Therefore, the probability of two sisters inheriting two identical, non-recombined maternal ChrXs is $2 \times 0.135^2 = 0.036$. This implies that, if two step-sisters share an identical haplotype A in addition to individual haplotypes B and C, the likelihood ratio of shared paternity vs. non-shared paternity equals

$$f(A) \cdot \frac{1}{2} \cdot 2f(B)f(C)/0.036 \cdot f(A) \cdot f(B) \cdot f(C) = 27.8$$

In other words, the probability of full sisterhood, assuming equal prior odds, cannot exceed $27.8/(1 + 27.8) = 0.965$.

Grandfather–grandson kinship is an excellent field for X marker usage. Passing the daughter generation, the grandpaternal X-chromosome normally underlies a recombination. However, typing haplotypes consisting of closely linked markers provides a high chance to indicate kinship. Figure 7.3 demonstrates a deficiency case that was solved using the Argus X-8 kit. Since the two questionable cousins

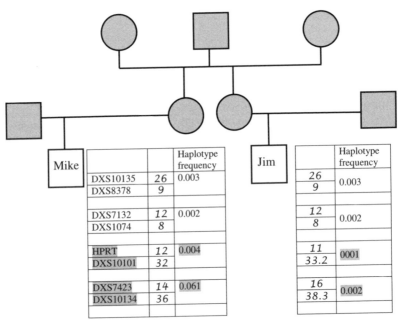

Mike			Haplotype frequency	Jim		Haplotype frequency
	DXS10135	26	0.003		26	0.003
	DXS8378	9			9	
	DXS7132	12	0.002		12	0.002
	DXS1074	8			8	
	HPRT	12	0.004		11	0001
	DXS10101	32			33.2	
	DXS7423	14	0.061		16	0.002
	DXS10134	36			38.3	

Figure 7.3 Complex kinship test using X-chromosomal STR haplotyping. Calculation was made to compute the likelihood that Mike and Jim are cousins (hypothesis H0) versus unrelated males (H1). The four STR clusters have been considered as unlinked markers. A likelihood ratio of 380 was obtained in favour of H0, which implies that, assuming equal prior odds, the probability of Mike being a cousin of Jim equals $380/(1 + 380) = 0.997$ (or 99.73%)

exhibit the ChrX haplotype with respect to two STR pairs (not shaded), we were able to demonstrate that putative cousins share a common grandfather.

7.7 Chromosome X–chromosome Y homologue markers

The ChrX–ChrY homologue non-recombining region are unique in the genome. Hosted on the ChrY this region can be transmitted only from father to son whereas the ChrX counterpart is inherited in the typical X-chromosomal mode. Doubtless, the most utilized locus of this type is the diallelic indel-polymorphism amelogenin, which is located at Yp11.1; Xp22.2 and is well-established for molecular gender assessment. Cali *et al.* (2002) described the locus DXYS156, which is located at Yp11.3 and Xq21.2-3. This marker is multi-allelic at both ChrX and ChrY. A Y-chromosome-specific nucleotide insertion in the duplicate STR allows males to be distinguished from females, as does the commonly used amelogenin system, but with the advantage that this locus, due to their multiple alleles, may contribute towards DNA fingerprinting of a sample. Yet another bonus is that both the X and the Y copies of DXYS156 have alleles specific to different parts of the world, offering separate estimates of maternal and paternal descent of that sample. It is of interest that some further markers of this kind have been published, namely DXYS241, DXYS265 and DXYS266 (Kotliarova *et al.*, 1999; Lee *et al.*, 2001). However, since they are dinucleotides they are sub-optimal in their use for forensic casework. Nevertheless, ChrX–ChrY homologue markers should attract more attention from the forensic community.

7.8 Chromosome X STR allele and haplotype distribution in different populations

The ChrX marker differences of allele distribution are marginal when nearly related populations are compared (Zarrabeitia *et al.*, 2006). However, when considered worldwide they may show noteworthy allele frequency differences (Lee *et al.*, 2004; Shin *et al.*, 2004, 2005; Tabbada *et al.*, 2005). Linkage disequilibrium (LD), or the non-random association of alleles, is not completely understood in the human genome. However, studies of LD between microsatellites and, more recently, between SNPs (Reich *et al.*, 2001) have provided new insights into the origin and history of human populations. For example, LD levels are much higher in non-African than in African populations (Reich *et al.*, 2001) and this may reflect a population bottleneck that is associated with the origin of non-African populations (Jorde *et al.*, 1998). In particular, LD between ChrX markers can indicate ethnic differences with high effectiveness (Edelmann *et al.*, 2002; Kaessmann *et al.*, 2002; Lee *et al.*, 2004). Thus, ChrX marker LD investigation may complement ChrY and mtDNA studies on human world migration.

7.9 Ethical considerations in ChrX marker testing

Detection of gonosomal aberrations and testicular feminization ChrX marker testing

In the most countries, such as Germany, forensic scientists follow the principle that forensic DNA testing should not disclose diseases or genetic risks. This principle fully complies with STR typing strategies using well-established autosomal STRs, such as CODIS markers, ChrY markers, mtDNA analysis and nearly all established ChrX markers. Both ChrY and mtDNA typing may reveal some general information as to a person's ethnic origin, however, this cannot be considered as an intervention into the person's privacy. In principle, the same applies to gender identification typing with ChrX and ChrY markers. However, chromosomal aberrations such as Klinefelter syndrome and Ullrich-Turner syndrome may be recognized when ChrX markers are used. This may be diagnosed when females show (virtual) homozygosity in all the ChrX STRs investigated. Furthermore, androgen insensitivity syndrome (known also under the alternative titles of 'testicular feminization syndrome, androgen receptor deficiency or dihydrotestosterone receptor deficiency') can be recognized when a person's female phenotype is linked not with a female genotype (XX) but with the male counterpart XY.

When gonosomal aberrations or instances of androgen insensitivity or XY gonadal dysgenesis are detected, ChrX typing is no longer a valid means of kinship testing. In any case, it appears worthwhile emphasizing that such findings, when inadvertently obtained during kinship testing, fall under the premise of confidentiality. Disease-relevant information should not be revealed to an affected individual unless they explicitly ask for it.

The HumARA trinucleotide repeat

Notwithstanding the fact that HumARA is one of the best-established forensic STR markers (Desmarais *et al.*, 1998), our group has recently announced that we no longer consider HumARA to be a suitable DNA marker in forensic casework (Szibor *et al.*, 2005a). From the very beginning of HumARA testing it has been known that, in contrast to all other forensic DNA markers, the HumARA CAG repeat is located in a coding region (androgen receptor gene, exon 1). This means that the repeat codes for a polyglutamine tract (La Spada *et al.*, 1991) proved that X-linked spinal and bulbar muscular atrophy (SBMA) is attributable to a mutation at this locus. This disease occurs at trinucleotide repeat lengths longer than 43. Apart from the SBMA disease, HumARA typing can detect a number of further health risks, e.g. increased risk of impaired spermatogenesis (Tut *et al.*, 1997), prostate cancer (Giovannucci *et al.*, 1997) and many more.

7.10 Concluding remarks

Since sex chromosome markers are especially efficient for solving deficiency cases, an increasing usage can be expected. A specific demand for kinship tests in which only remote relatives are available for testing can be expected to arise, particularly from the need to rejoin families in the context of war and worldwide migration. Here, ChrX marker testing may also prove helpful. Furthermore, the proportion of non-marital children is constantly increasing in modern industrial societies and, for example, accounts for approximately 50% of all births in the eastern federal states of Germany. In many of these instances, paternity may be disputed at some stage and, when the putative father dies early or unexpectedly, the need for a paternity test may only be recognized after the interment.

The present chapter was intended to highlight the potential of ChrX STRs for solving some of the above problems.

7.11 References

Adam, A., Sheba, C., Sanger, R., *et al.* (1963) Data for X-mapping calculations, Israeli families tested for Xg, g-6-pd and for colour vision. *Ann. Hum. Genet.* **26**: 187–194.

Adam, A., Tippett, P., Gavin, J., *et al.* (1967) The linkage relation of Xg to g-6-pd in Israelis: the evidence of a second series of families. *Ann. Hum. Genet.* **30**: 211–218.

Asamura, H., Sakai, H., Kobayashi, K., *et al.* (2006) MiniX-STR multiplex system population study in Japan and application to degraded DNA analysis. *Int. J. Legal Med.* **120**: 174–181.

Askov, M., Kutlar, A., Kutlar, F., *et al.* (1985) Survey on haemoglobin variants, beta thalassaemia, glucose-6-phosphate dehydrogenase deficiency, and haptoglobin types in Turks from western Thrace. *J. Med. Genet.* **22**: 288–290.

Augustin, C., Cichy, R., Hering, S., *et al.* (2006) Forensic evaluation of three closely linked STR markers in a 13 kb region at Xp11.23. *Int. Congr. Ser.* **1239**: 311–314.

Banchs, I., Bosch, A., Guimera, J., *et al.* (1994) New alleles at microsatellite loci in CEPH families mainly arise from somatic mutations in the lymphoblastoid cell lines. *Hum. Mutat.* **3**: 365–372.

Bär, W., Brinkmann, B., Budowle, B., *et al.* (1997) DNA recommendations. Further report of the DNA Commission of the ISFH regarding the use of short tandem repeat systems. *Int. J. Legal Med.* **110**: 175–176.

Barr, M.L. and Bertram, E.G. (1949) A morphological distinction between neurons of the male and female, and the behaviour of the nucleolar satellite during accelerated nucleoprotein synthesis. *Nature* **163**: 676–677.

Barr, M.L. and Carr, D.H. (1962) Correlations between sex chromatin and sex chromosomes. *Acta Cytol.* **6**: 34–45.

Bein, G., Driller, B., Schurmann, M., *et al.* (1998) Pseudo-exclusion from paternity due to maternal uniparental disomy 16. *Int. J. Legal Med.* **111**: 328–330.

Beutler, E. (1969) Electrophoresis of phosphoglycerate kinase. *Biochem. Genet.* **3**: 189–195.

Bojesen, A., Juul, S. and Gravholt, C.H. (2003) Prenatal and postnatal prevalence of Klinefelter syndrome: a national registry study. *J. Clin. Endocrinol. Metab.* **88**: 622–626.

Brinkmann, B. (1998) Overview of PCR-based systems in identity testing. *Methods Mol. Biol.* **98**: 105–119.

Bucher, K.D. and Elston, R.C. (1975) Letter: Estimation of nonpaternity for X-linked trait. *Am. J. Hum. Genet.* **27**: 689–690.

Cali, F., Forster, P., Kersting, C., *et al.* (2002) DXYS156: a multi-purpose short tandem repeat locus for determination of sex, paternal and maternal geographic origins and DNA fingerprinting. *Int. J. Legal Med.* **116**: 133–138.

Chen, S.H., Malcolm, L.A., Yoshida, A., *et al.* (1971) Phosphoglycerate kinase: an X-linked polymorphism in man. *Am. J. Hum. Genet.* **23**: 87–91.

Clement-Jones, M., Schiller, S., Rao, E., *et al.* (2000) The short stature homeobox gene SHOX is involved in skeletal abnormalities in Turner syndrome. *Hum. Mol. Genet.* **9**: 695–702.

Davidson, W.M. and Smith, D.R. (1954) A morphological sex difference in the polymorphonuclear neutrophil leucocytes. *Br. Med. J.* **4878**: 6–7.

Davies, K.E., Speer, A., Herrmann, F., *et al.* (1985) Human X chromosome markers and Duchenne muscular dystrophy. *Nucleic Acids Res.* **13**: 3419–3426.

Davies, S.H., Gavin, J., Goldsmith, K.L., *et al.* (1963) The linkage relations of hemophilia A and hemophilia B (Christmas Disease) to the Xg blood group system. *Am. J. Hum. Genet.* **15**: 481–492.

de Knijff, P., Kayser, M., Caglia, A., *et al.* (1997) Chromosome Y microsatellites: population genetic and evolutionary aspects. *Int. J. Legal Med.* **110**: 134–149.

Deloukas, P., Schuler, G.D., Gyapay, G., *et al.* (1998) A physical map of 30,000 human genes. *Science* **282**: 744–746.

Desmarais, D., Zhong, Y., Chakraborty, R., *et al.* (1998) Development of a highly polymorphic STR marker for identity testing purposes at the human androgen receptor gene (HUMARA). *J. Forensic Sci.* **43**: 1046–1049.

Duma, A. and Boskovski, K. (1977) [Forensic significance of Barr bodies]. *God. Zb. Med. Fak. Skopje.* **23**: 461–466.

Edelmann, J., Deichsel, D., Plate, I., *et al.* (2003) Validation of the X-chromosomal STR DXS6809. *Int. J. Legal Med.* **117**: 241–244.

Edelmann, J., Hering, S., Kuhlisch, E., *et al.* (2002) Validation of the STR DXS7424 and the linkage situation on the X-chromosome. *Forensic Sci. Int.* **125**: 217–222.

Edelmann, J., Hering, S., Michael, M., *et al.* (2001) 16 X-chromosome STR loci frequency data from a German population. *Forensic Sci. Int.* **124**: 215–218.

Edelmann, J., Lessig, R., Willenberg, A., *et al.* (2006) Forensic validation of the X-chromosomal STR-markers GATA165B12, GATA164A09, DXS9908 and DXS7127 in German population. *Int. Congr. Ser.* **1239**: 298–300.

Edelmann, J. and Szibor, R. (2005) Validation of the X-linked STR DXS6801. *Forensic Sci. Int.* **148**: 219–220.

Edelmann, J. and Szibor, R. (1999) Validation of the HumDXS6807 short tandem repeat polymorphism for forensic application. *Electrophoresis* **20**: 2844–2846.

Edelmann, J. and Szibor, R. (2001) DXS101: a highly polymorphic X-linked STR. *Int. J. Legal Med.* **114**: 301–304.

Edelmann, J. and Szibor, R. (2003) The X-linked STRs DXS7130 and DXS6803. *Forensic Sci. Int.* **136**: 73–75.

Edwards, A., Hammond, H.A., Jin, L., *et al.* (1992) Genetic variation at five trimeric and tetrameric tandem repeat loci in four human population groups. *Genomics* **12**: 241–253.

Ellis, N.A., Tippett, P., Petty, A., *et al.* (1994) PBDX is the XG blood group gene. *Nat. Genet.* **8**: 285–290.

Giovannucci, E., Stampfer, M.J., Krithivas, K., *et al.* (1997) The CAG repeat within the androgen receptor gene and its relationship to prostate cancer. *Proc. Natl. Acad. Sci. USA* **94**: 3320–3323.

Given, B.W. (1976) Sex-chromatin bodies in penile washings as an indicator of recent coitus. *J. Forensic Sci.* **21**: 381–386.

Haldane, J.B.S. (1919) The combination of linkage values and the calculation of distances between the loci for linked factors. *J. Genet.* **8**: 299–309.

Hearne, C.M. and Todd, J.A. (1991) Tetranucleotide repeat polymorphism at the HPRT locus. *Nucleic Acids Res.* **19**: 5450.

Hellmann, A., Rohleder, U., Schmitter, H., *et al.* (2001) STR typing of human telogen hairs – a new approach. *Int. J. Legal Med.* **114**: 269–273.

Hering, S., Augustin, C., Edelmann, J., *et al.* (2006) DXS10079, DXS10074 and DXS10075 are STRs located within a 280-kb region of Xq12 and provide stable haplotypes useful for complex kinship cases. *Int. J. Legal Med.* **120**: 337–345.

Hering, S., Brundirs, N., Kuhlisch, E., *et al.* (2004) DXS10011: studies on structure, allele distribution in three populations and genetic linkage to further q-telomeric chromosome X markers. *Int. J. Legal Med.* **118**: 313–319.

Hering, S., Kuhlisch, E. and Szibor, R. (2001) Development of the X-linked tetrameric microsatellite marker HumDXS6789 for forensic purposes. *Forensic Sci. Int.* **119**: 42–46.

Hering, S. and Szibor, R. (2000) Development of the X-linked tetrameric microsatellite marker DXS9898 for forensic purposes. *J. Forensic Sci.* **45**: 929–931.

Jobling, M.A., Pandya, A. and Tyler-Smith, C. (1997) The Y chromosome in forensic analysis and paternity testing. *Int. J. Legal Med.* **110**: 118–124.

Jorde, L.B., Bamshad, M. and Rogers, A.R. (1998) Using mitochondrial and nuclear DNA markers to reconstruct human evolution. *Bioessays* **20**: 126–136.

Kaessmann, H., Zollner, S., Gustafsson, A.C., *et al.* (2002) Extensive linkage disequilibrium in small human populations in Eurasia. *Am. J. Hum. Genet.* **70**: 673–685.

Karolchik, D., Baertsch, R., Diekhans, M., *et al.* (2003) The UCSC Genome Browser Database. *Nucleic Acids Res.* **31**: 51–54.

Kayser, M., Caglia, A., Corach, D., *et al.* (1997a) Evaluation of Y-chromosomal STRs: a multicenter study. *Int. J. Legal Med.* **110**: 125–133, 141–149.

Kayser, M., de Knijff, P., Dieltjes, P., *et al.* (1997b) Applications of microsatellite-based Y chromosome haplotyping. *Electrophoresis* **18**: 1602–1607.

Kishida, T. and Tamaki, Y. (1997) Japanese population data on X-chromosomal STR locus AR. *Nippon Hoigaku Zasshi* **51**: 376–379.

Kishida, T., Wang, W., Fukuda, M., *et al.* (1997) Duplex PCR of the Y-27H39 and HPRT loci with reference to Japanese population data on the HPRT locus. *Nippon Hoigaku Zasshi* **51**: 67–69.

Kotliarova, S.E., Toda, T., Takenaka, O., *et al.* (1999) Novel (CA)n marker DXYS241 on the nonrecombinant part of the human Y chromosome. *Hum. Biol.* **71**: 261–275.

Krüger, J., Fuhrmann, W., Lichte, K.H., *et al.* (1968) [On the utilization of erythrocyte acid phosphatase polymorphism in paternity evaluation]. *Dtsch. Z. Gesamte Gerichtl. Med.* 64: 127–146.

La Spada, A.R., Wilson, E.M., Lubahn, D.B., *et al.* (1991) Androgen receptor gene mutations in X-linked spinal and bulbar muscular atrophy. *Nature* 352: 77–79.

Lalloz, M.R., McVey, J.H., Pattinson, J.K., *et al.* (1991) Haemophilia A diagnosis by analysis of a hypervariable dinucleotide repeat within the factor VIII gene. *Lancet* 338: 207–211.

Lalloz, M.R., Schwaab, R., McVey, J.H., *et al.* (1994) Haemophilia A diagnosis by simultaneous analysis of two variable dinucleotide tandem repeats within the factor VIII gene. *Br. J. Haematol* 86: 804–809.

Lambropoulos, A.F., Frangoulides, E., Kotsis, A., *et al.* (1995) Rapid typing of 4 VNTR loci, 3'ApoB, MCT118, St14 and YNZ22 by the polymerase chain reaction of a Greek sample. *Cell. Mol. Biol.* 41: 699–702.

Lee, H.Y., Park, M.J., Jeong, C.K., *et al.* (2004) Genetic characteristics and population study of 4 X-chromosomal STRs in Koreans: evidence for a null allele at DXS9898. *Int. J. Legal Med.* 118: 355–360.

Lee, J., Kotliarova, S.E., Ewis, A.A., *et al.* (2001) Y chromosome compound haplotypes with the microsatellite markers DXYS265, DXYS266, and DXYS241. *J. Hum. Genet.* 46: 80–84.

Lewis, F.J., Froland, A., Sanger, R., *et al.* (1964) Source of the X chromosomes in two XXXXY males. *Lancet* 14: 589.

Lyon, M.F. (1961) Gene action in the X-chromosome of the mouse (*Mus musculus* L.). *Nature* 190: 372–373.

Mahtani, M.M. and Willard, H.F. (1993) A polymorphic X-linked tetranucleotide repeat locus displaying a high rate of new mutation: implications for mechanisms of mutation at short tandem repeat loci. *Hum. Mol. Genet.* 2: 431–437.

Mann, J.D., Cahan, A., Gelb, A.G., *et al.* (1962) A sex-linked blood group. *Lancet* 1: 8–10.

McNeil, J.A., Smith, K.P., Hall, L.L., *et al.* (2006) Word frequency analysis reveals enrichment of dinucleotide repeats on the human X chromosome and [GATA]n in the X escape region. *Genome Res.* 16: 477–484.

Meissner, C., Bruse, P., Mueller, E., *et al.* (2006) A new sensitive short pentaplex (ShoP) PCR for typing of degraded DNA. *Forensic Sci. Int.* [Epub ahead of print] doi:10.1016/j.forsciint.2006.04.014.

Mudd, J.L. (1984) The determination of sex from forcibly removed hairs. *J. Forensic Sci.* 29: 1072–1080.

Nagaraja, R., MacMillan, S., Jones, C., *et al.* (1998) Integrated YAC/STS physical and genetic map of 22.5 Mb of human Xq24-q26 at 56-kb inter-STS resolution. *Genomics* 52: 247–266.

Nei, M. and Roychoudhury, A.K. (1974) Sampling variances of heterozygosity and genetic distance. *Genetics* 76: 379–390.

Oberle, I., Camerino, G., Heilig, R., *et al.* (1985) Genetic screening for hemophilia A (classic hemophilia) with a polymorphic DNA probe. *N. Engl. J. Med.* 312: 682–686.

Ott, J. (1991) *Analysis of Human Genetic Linkage.* Johns Hopkins University Press, Baltimore.

Poetsch, M., Petersmann, H., Repenning, A., *et al.* (2005a) Development of two penta-plex systems with X-chromosomal STR loci and their allele frequencies in a northeast German population. *Forensic Sci. Int.* **155**: 71–76.

Poetsch, M., Repenning, A., Lignitz, E., *et al.* (2005b) DXS6797 contains two STRs which can be easily haplotyped in both sexes. *Int. J. Legal Med.* **120**: 61–66.

Reich, D.E., Cargill, M., Bolk, S., *et al.* (2001) Linkage disequilibrium in the human genome. *Nature* **411**: 199–204.

Roewer, L., Krawczak, M., Willuweit, S., *et al.* (2001) Online reference database of European Y-chromosomal short tandem repeat (STR) haplotypes. *Forensic Sci. Int.* **118**: 106–113.

Ross, M.T., Grafham, D.V. and Coffey, A.J. (2005) The DNA sequence of the human X chromosome. *Nature* **434**: 325–337.

Roychoudhury, A.K. and Nei, M. (1988) *Human Polymorphic Genes: World Distribution*, Oxford University Press, New York.

Sanger, R., Tippett, P. and Gavin, J. (1971) Xg groups and sex abnormalities in people of northern European ancestry. *J. Med. Genet.* **8**: 417–426.

Sanger, R., Tippett, P., Gavin, J., *et al.* (1977) Xg groups and sex chromosome abnor-malities in people of northern European ancestry: an addendum. *J. Med. Genet.* **14**: 210–211.

Schmidtke, J., Kuhnau, W., Wand, D., *et al.* (2004) Prenatal exclusion without involving the putative fathers of an incestuous father-daughter parenthood. *Prenat. Diagn.* **24**: 662–664.

Shin, K.J., Kwon, B.K., Lee, S.S., *et al.* (2004) Five highly informative X-chromosomal STRs in Koreans. *Int. J. Legal Med.* **118**: 37–40.

Shin, S.H., Yu, J.S., Park, S.W., *et al.* (2005) Genetic analysis of 18 X-linked short tandem repeat markers in Korean population. *Forensic Sci. Int.* **147**: 35–41.

Sleddens, H.F., Oostra, B.A., Brinkmann, A.O., *et al.* (1992) Trinucleotide repeat poly-morphism in the androgen receptor gene (AR). *Nucleic Acids Res.* **20**: 1427.

Subramanian, S., Mishra, R.K. and Singh, L. (2003) Genome-wide analysis of micros-atellite repeats in humans: their abundance and density in specific genomic regions. *Genome Biol.* **4**: R13.

Szibor, R., Edelmann, J., Zarrabeitia, M.T., *et al.* (2003) Sequence structure and popula-tion data of the X-linked markers DXS7423 and DXS8377 – clarification of conflict-ing statements published by two working groups. *Forensic Sci. Int.* **134**: 72–73.

Szibor, R., Hering, S. and Edelmann, J. (2005a) The HumARA genotype is linked to spinal and bulbar muscular dystrophy and some further disease risks and should no longer be used as a DNA marker for forensic purposes. *Int. J. Legal Med.* **119**: 179–180.

Szibor, R., Hering, S., Kuhlisch, E., *et al.* (2005b) Haplotyping of STR cluster DXS6801-DXS6809-DXS6789 on Xq21 provides a powerful tool for kinship testing. *Int. J. Legal Med.* **119**: 363–369.

Szibor, R., Lautsch, S., Plate, I., *et al.* (2000) Population data on the X chromosome short tandem repeat locus HumHPRTB in two regions of Germany. *J. Forensic Sci.* **45**: 231–233.

Tabbada, K.A., De Ungria, M.C., Faustino, L.P., *et al.* (2005) Development of a penta-plex X-chromosomal short tandem repeat typing system and population genetic studies. *Forensic Sci. Int.* **154**: 173–180.

Tippett, P. and Ellis, N.A. (1998) The Xg blood group system: a review. *Transfus. Med. Rev.* **12**: 233–257.

Tröger, H.D., Liebhardt, E. and Eisenmenger, W. (1976) [Who smoked the cigarette? Determination of the male nuclear sex in mouth mucosa cells]. *Beitr. Gerichtl. Med.* **34**: 207–209.

Tut, T.G., Ghadessy, F.J., Trifiro, M.A., *et al.* (1997) Long polyglutamine tracts in the androgen receptor are associated with reduced trans-activation, impaired sperm production, and male infertility. *J. Clin. Endocrinol. Metab.* **82**: 3777–3782.

Urquhart, A., Kimpton, C.P., Downes, T.J., *et al.* (1994) Variation in short tandem repeat sequences – a survey of twelve microsatellite loci for use as forensic identification markers. *Int. J. Legal Med.* **107**: 13–20.

Watanabe, G., Umetsu, K., Yuasa, I., *et al.* (2000) DXS10011: a hypervariable tetranucleotide STR polymorphism on the X chromosome. *Int. J. Legal Med.* **113**: 249–250.

Wegener, R., Wkirich, V., Dauber, E.M., *et al.* (2006) Mother-child exclusion due to paternal uniparental disomy 6. *Int. J. Legal Med.* **120**: 282–285.

Went, L.N., De Groot, W.P., Sanger, R., *et al.* (1969) X-linked ichthyosis: linkage relationship with the Xg blood groups and other studies in a large Dutch kindred. *Ann. Hum. Genet.* **32**: 333–345.

Wieacker, P.F., Knoke, I. and Jakubiczka, S. (1998) Clinical and molecular aspects of androgen receptor defects. *Exp. Clin. Endocrinol. Diabetes* **106**: 446–453.

Wiegand, P. and Kleiber, M. (2001) Less is more – length reduction of STR amplicons using redesigned primers. *Int. J. Legal Med.* **114**: 285–287.

Xiao, F.X., Gilissen, A., Cassiman, J.J., *et al.* (1998) Quadruplex fluorescent STR typing system (HUMVWA, HUMTH01, D21S11 and HPRT) with sequence-defined allelic ladders identification of a new allele at D21S11. *Forensic Sci. Int.* **94**: 39–46.

Yoshida, A., Giblett, E.R. and Malcolm, L.A. (1973) Heterogeneous distribution of glucose-6-phosphate dehydrogenase variants with enzyme deficiency in the Markham Valley Area of New Guinea. *Ann. Hum. Genet.* **37**: 145–150.

Yun, W.M. and Yun, S.G. (1996) Analysis of the VNTR locus DXS52 by the Amp-FLP technique. *J. Forensic Sci.* **41**: 859–861.

Zarrabeitia, M.T., Alonso, A., Martin, J., *et al.* (2006) Study of six X-linked tetranucleotide microsatellites: population data from five Spanish regions. *Int. J. Legal Med.* **120**: 147–150.

8

Mitochondrial analysis in forensic science

Hirokazu Matsuda and Nobuhiro Yukawa

8.1 Introduction

Mammalian mitochondrial DNA (mtDNA) is a small 16.5 kbp circular genome. Human mtDNA was found to contain two adjacent, highly polymorphic regions, which are designated hypervariable regions I and II (HV-I and HV-II). The most common polymorphisms are nucleotide substitutions (transitions and transversions) followed by deletions and insertions (Butler and Levin, 1998; Meyer et al., 1999). Polymerase chain reaction (PCR)-amplification and subsequent sequencing of these two polymorphic regions is currently referred to as 'mtDNA typing' and the sequences obtained by this technique as 'mtDNA types' (Holland and Parsons, 1999; Carracedo et al., 2000).

Short tandem repeat (STR) typing, an established forensic typing method based upon nuclear DNA (nDNA) polymorphisms, is highly effective when DNA is of sufficient amount and quality (Moretti et al., 2001). Frequently, due to inadequate quantity or degradation of the sample into small fragments, DNA extracted from forensic samples is of limited use. Samples notorious for unsuccessful STR typing include old bones, teeth and hair, particularly hair that has been shed, as these samples generally lack detectable nDNA. Even blood and body fluid samples, which are frequently used for STR typing, occasionally give unsuccessful results due to sample aging or decomposition. For these types of samples, mtDNA typing has proven to be more successful than STR typing, though the individual discrimination power of mtDNA typing is in general lower than with STR typing.

Molecular Forensics. Edited by Ralph Rapley and David Whitehouse
Copyright 2007 by John Wiley & Sons, Ltd.

This chapter describes the relevance of mtDNA biology to forensics, mtDNA typing and two additional topics of forensic interest: species identification through mtDNA analysis and the attempt to increase individual discrimination power by typing single nucleotide polymorphisms (SNPs). Detailed reviews of mtDNA forensic applications can be found in Butler and Levin (1998), Holland and Parsons (1999) and Budowle *et al.*, (2003).

8.2   Mitochondrial DNA (mtDNA) biology

Origins of mitochondria and mtDNA

The production of ATP by the process of oxidative phosphorylation is the principal function of the mitochondrion, which is an organelle of eukaryotes (fungi, plants and animals). Endosymbiosis may explain how mitochondria came to be incorporated within eukaryotic cells (Embley and Martin, 2006). While not the only endosymbiotic theory describing the evolution of mitochondria, the hydrogen hypothesis (Gray *et al.*, 1999) postulates that small ancient bacteria capable of producing hydrogen (H_2) were integrated into and survived within larger H_2-consuming bacteria. Through this symbiotic relationship, primordial mitochondria evolved from the small H_2-producing bacteria and acquired the ability to conduct oxidative phosphorylation. Eventually, the eukaryotic (enveloped) nucleus was derived from genes of the larger bacteria. Over evolutionary time, most of the initial mitochondrial genes (genes from the small bacteria) were transferred into the eukaryotic nucleus. Genes escaping transfer eventually developed into the present-day mtDNA genome.

High copy number per cell

Most mammalian cells contain a dozen to hundreds of mitochondria, though it was found that mitochondria are not static units. They are capable of dynamically fusing with and separating from each other to form a single functionally complex network structure (Hayashi *et al.*, 1994). Each mitochondrial unit (mitochondrion) contains a few to a dozen copies of mtDNA genomes. One somatic cell may contain hundreds to more than a thousand copies of identical mtDNA genomes, with larger numbers being found in tissues that demand a greater amount of oxygen, such as the brain and skeletal muscles (Shuster *et al.*, 1988; Tully and Levin, 2000). This is in contrast with the two copies of nDNA genomes per somatic diploid cell. The relative abundance of mtDNA imparts a correspondingly higher degree of recovery from forensic samples, and is one of the principal reasons why mtDNA typing achieves a higher degree of sensitivity than those obtained by STR typing.

Maternal inheritance and recombination rate

In mammals, each somatic diploid cell has two copies of the nDNA genomes, with one inherited from each parent. On the other hand, the progeny inherits its mtDNA directly from the mother (Giles *et al.*, 1980). Mechanisms behind the maternal inheritance of mtDNA include the reduction of paternal (spermatic) mtDNA during spermatogenesis, the simple dilution of spermatic mtDNA at fertilization (an overwhelming copy number of oocyte mtDNA relative to spermatic mtDNA) and ubiquitin-mediated proteolysis of spermatic mitochondria and the active digestion of spermatic mtDNA within a fertilized ovum. (Kaneda *et al.*, 1995; Hershko and Ciechanover, 1998; Nishimura *et al.*, 2006). Due to these numerous safeguards, the paternal mtDNA that enters an oocyte becomes undetectable after the fertilized egg undergoes its first mitotic division.

Whether mammalian mtDNA undergoes recombination or not is a longstanding question. In recent experiments using human somatic hybrid cells and mice carrying two different mtDNA, only three of 318 clones of mtDNA purified from mouse tissues corresponded to recombinant mtDNA, whereas no recombinants were found in human somatic hybrid cells. These results strongly suggest that recombination can occur within mammalian cells but at a very low frequency or at an operationally undetectable level. This implies that recombinant mtDNA observed in mice might be gene conversion products resulting from the repair of damaged mtDNA molecules (Sato *et al.*, 2005)

Maternal inheritance and a very low recombination rate means that mtDNA genomes are essentially clonal copies of the mother's mtDNA genomes (provided that the sequence of all mtDNA molecules within an oocyte are the same, the recombinant mtDNA structure will also be the same). From a forensic viewpoint, maternal inheritance can be a helpful tool in the identification of a body or the remains of a missing person. The missing person's biological mother, siblings and maternal relatives all have the same mtDNA sequence, with few exceptions resulting from heteroplasmy as described later. Therefore, biological samples (e.g. blood, buccal swabs) taken from these individuals can provide reference samples for the identification of the missing individual. On the other hand, since a progeny's mtDNA contains no paternal information, mtDNA typing cannot be used for paternity tests.

High mutation rates

In addition to maternal inheritance, mtDNA and nDNA mutation rates differ. The low fidelity of mtDNA polymerase, the lack of protective histone proteins and a less effective repair system lead to a higher base substitution rate in mtDNA (Pinz and Bogenhagen, 1998). Within mtDNA, HV-I and HV-II appear to evolve quickly, with their mutation rates being about 5–10 times higher than that of nDNA (Budowle *et al.*, 2003). Together with maternal inheritance, higher mutation rates have made mtDNA typing an interesting tool for human population genetics and evolutionary studies (Stoneking, 1994). Additionally, the somatic accumulation of mtDNA mutations has been proposed to play a role in human aging (Michikawa *et al.*, 1999).

Structure of mtDNA

The complete sequence of human mtDNA was first determined in the laboratory of Frederick Sanger in Cambridge, England (Anderson *et al.*, 1981). This human mtDNA sequence, also called Anderson reference, is designated as the Cambridge Reference Sequence (CRS). Following the sequencing of human mtDNA, mtDNA sequences of animals were determined. A comparison of sequences revealed that the gross structure and genetic arrangements are remarkably conserved among mammalian species. Human mtDNA is a double-stranded circular molecule 16 569 bp in length (Figure 8.1). Based upon differences in buoyant density, the strands are termed the heavy strand (H-strand) and the light strand (L-strand). The H-strand is rich in purines (A, G), whereas within the L-strand, pyrimidines (T, C) dominate. When metabolically active cells are observed by electron microscopy, a large population of mtDNA appears to contain a short three-strand structure. This structure represents an initial stage of replication and is called a displacement loop (D-loop) (Taanman, 1999; Brown *et al.*, 2006).

According to the numbering system offered by the CRS, the initial position '1' was arbitrarily assigned near the middle of the control region. The base number then increases in the (5'→3') direction on the L-strand, and, because of its circular nature, the final position '16 569' is located next to '1' (Figure 8.1). The CRS was revised and termed the 'rCRS' (Andrews *et al.*, 1999). By this revision, the original CRS, which had been determined from a single individual, was found to contain several rare polymorphisms. This discovery emphasized that the CRS (rCRS) cannot be regarded as the 'authentic' sequence but should be used as a 'reference' sequence to facilitate the comparison among sequences cited in the literature and those determined from samples. It is also noted that, because of the same gross structure and genetic arrangements among mammalians, animal mtDNAs can be numbered using CRS (rCRS).

Functionally, mammalian and hence human mtDNA is divided into coding and control regions (Holland and Parsons, 1999; Taanman, 1999). The coding region contains 37 intron-less genes encoding 2 ribosomal RNAs (12S and 16S rRNAs), 22 transfer RNAs (tRNAs) and 13 protein enzymes. The 22 tRNAs are the minimum set required for the translation of mtDNA, and all of the 13 proteins are involved in the process of oxidative phosphorylation.

The control region, which corresponds to the D-loop, is bound by the genes for tRNA[phe] and tRNA[pro] (Figure 8.1). The length is 1122 bp (CRS) and may

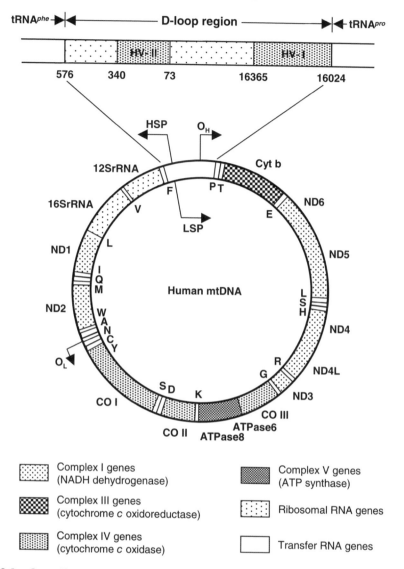

Figure 8.1 Genetic map of human mtDNA and expanded diagram of the D-loop region. The outer circle represents the H-strand, containing the majority of the genes; the inner circle represents the L-strand. The origins of replication for H-strand (O_H) and L-strand (O_L), and the promoters for H-strand (HSP) and L-strand (LSP), are shown by the arrows. The 22 tRNA genes are depicted by the single-letter code of the amino acid (isoacceptors for serine and leucine are distinguished by their codon sequence). The D-loop region diagram shows flanking tRNAs and location of hypervariable regions HV-I and HV-II; numbering system follows that of the CRS

vary by one or more bases owing to deletions, insertions or repetition. The control region contains the binding sites for the major promoters of transcription and the origin of H-strand replication (OH) (Taanman, 1999). This region also contains the HV-I and HV-II polymorphic regions. The HV-I ranges from position 16 024 to 16 365, and HV-II ranges from position 73 to 340. These hypervariable regions represent ongoing mutational hotspots rather than the remnant of previously incorporated or fixed mutations (Stoneking, 2000).

Heteroplasmy

Reflecting the clonal replication of maternal mtDNA, all copies of mtDNA genomes are identical (homoplasmic) as a rule. Due to the high copy number of mtDNA, however, a mutation in some of the mtDNA results in a mixture of variant mtDNA genomes, a condition known as heteroplasmy (Holt *et al.*, 1990). Heteroplasmy is operationally defined as the presence of two or more subpopulations (types) of mtDNA genomes within a mitochondrion, cell, tissue, organ or individual, and may be observed in several ways, such as two or more mtDNA types in one tissue sample, and one mtDNA type in one tissue sample and a different mtDNA type in another sample (Budowle *et al.*, 2003). On occasion, children of a heteroplasmic mother may be homoplasmic. This occurs through the inheritance of the mother's predominant mtDNA type (or one of the predominant types) due to a 'bottleneck' mechanism that segregates minor mtDNA types (Holland and Parsons, 1999).

Heteroplasmy is most often observed in hair samples because genetic drift is allowed to operate and bottlenecks are created due to a hair follicle's semiclonal nature (Budowle *et al.*, 2003). One disadvantage of using mtDNA for forensic individual identification is the possibility that the occurrence of heteroplasmy will confuse the interpretation of the results and potentially lead to an erroneous exclusion rather than a match. However, the presence of heteroplasmy can also increase the power of the match when it is present in both the unknown and reference samples.

8.3 Identification of individuals (mtDNA typing)

Procedures and interpretation of results

Due to maternal inheritance and the very low rate of recombination, all copies of mtDNA are generally identical (homoplasmic). In other words, the mtDNA genome is haploid and its sequence is treated as a single locus (haplotype). Thus, two hypervariable regions (HV-I and HV-II) of mtDNA can be determined by direct sequencing after PCR-amplification of the regions. Even for heteroplasmic mtDNA containing two mtDNA types, the sequences can often be determined by direct sequencing because there are usually only one or at most two base differences between the two types.

Mitochondrial DNA typing starts with the extraction of total genomic DNA from samples, and then PCR-amplification is performed on HV-I and HV-II. Forensic investigators thought that the size of each region (~400 bp) was too large to amplify simply by using one primer that targets the entire HV-1 region and another primer that targets the entire HV-II region. As it is difficult to amplify large fragment sizes in aged or decomposed samples, they reasoned that reducing fragment sizes in such samples could allow for PCR-amplification to be performed. They therefore targeted amplicons ~250 bp in length, including the primer binding sites, with a total of four overlapping primer sets for HV-I and HV-II (Figure 8.2) (Stoneking *et al.*, 1991; Wilson *et al.*, 1995). The PCR-amplified products were then sequenced using a fluorescent automated sequencing system, and the sequencing information was confirmed by analysis of both forward and reverse DNA strands. The sequence of the unknown (evidence) sample was displayed as the L-strand sequence, and nucleotide differences between the evidence sample and CRS (rCRS) were noted. For example, an mtDNA type of '263G, 315.1C' indicates that an adenine (A) at position 263 in the CRS was substituted by a guanine residue (G), that there was an insertion of cytosine (C) between position 315 and 316 and that the remaining sequence of this mtDNA type was the same as that of the CRS.

In order to evaluate the sequencing results from evidence and reference samples, useful interpretation guidelines have been provided by the DNA Commission of the International Society for Forensic Genetics (Carracedo *et al.*, 2000) as follows:

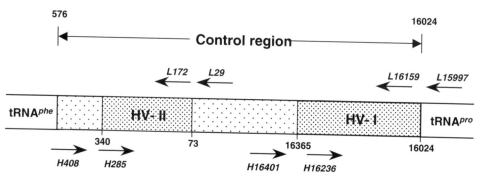

Figure 8.2 Diagram of the human mtDNA control region. Positions of the primers are indicated by arrows. For analysis of aged samples, two sets of PCR primers are used to amplify overlapping fragments ~250 bp long: *L15997/H16236* and *L16159/H16401* for HV-I and *L29/H285* and *L172/H408* for HV-II

If the sequences are unequivocally different, then the samples can be excluded as having originated from the same source. If the sequences are the same, then the reference and evidence samples cannot be excluded as potentially being from the same source. In cases where the same heteroplasmy is observed in both the known and unknown samples, its presence may increase the strength of the evidence. If heteroplasmy is observed in the questioned sample but not in the known sample, a common maternal lineage cannot be excluded. If the two samples differ by a single nucleotide, and there is no indication of heteroplasmy, the interpretation may be that the results are inconclusive. However, a one-nucleotide difference between two samples, on occasions, may provide evidence against the samples either originating from the same source or having the same maternal lineage; in particular, where both samples are a tissue such as blood, a single nucleotide difference points towards exclusion of a common maternal origin. The source of the tissue being investigated should be taken into consideration, because differences in mtDNA sequences due to mutations seem to be more likely between e.g. hair and blood than between two blood samples taken from the same individual.

Technical difficulty due to C-stretch

Both HV-I and HV-II of the human mtDNA contain homopolymeric tract of cytosines (C) (i.e., C-stretch) (Carracedo *et al.*, 2000). In HV-I, the most common sequence between positions 16 184 and 16 193 is ••CCCCCTCCCC•• (the position of thymine 'T' is at 16 189). When the T undergoes mutational transition to C, length heteroplasmy (polymorphism of the number of repeated cytosine residues) supervenes. A similar homopolymeric region resides in HV-II at positions 303–315 (Stewart *et al.*, 2001). These length heteroplasmies are believed to be supervened by an additional increase or decrease of cytosine residues through the replication slippage mechanism, which has been observed in poly G:C tracts (Hauswirth *et al.*, 1984). In general, direct sequencing of DNA from these samples exhibits reading frame shifts and thereby both the exact number of cytosine residues of the C-stretch and sequences beyond the C-stretch cannot be determined (Tully and Levin, 2000). The additional reactions using primers that sit on the C-stretch provide sequence information from both strands following the C-stretch but the exact number of cytosine residues in the C-stretch still cannot be accurately determined.

8.4   Topics of forensic interest

Species identifications

Samples subjected to mtDNA typing are not necessarily derived from humans, but usually this does not impose much trouble because the sequencing of HV-I and HV-II not only provides individual identification but also confirms whether the sample is of human origin. However, when amplification of HV-I and HV-II fails, species identification of DNA extracted from a sample may be useful in order to ascertain whether failure in amplification is due to the sample being non-human in origin (e.g. plants and animals) or whether it resulted from other causes (e.g. an insufficient amount of DNA and/or decomposition).

Cytochrome *b* (cyt *b*) is one component of the electron transport (respiratory chain) enzyme complexes involved in oxidative phosphorylation. In many types of species, including plants and animals, the cyt *b* gene exists in mtDNA (Figure 8.1). In view of the endosymbiosis hypothesis, the cyt *b* gene escaped incorporation by the nucleus and accumulated mutations over evolutionary time, leading to the present cyt *b* gene sequence differences among species. In other words, cyt *b* genes are species-specific and have been used for species determination in phylogenetic and forensic investigations (Kocher *et al.*, 1989; Parson *et al.*, 2000). In these investigations, a portion of the cyt *b* gene is amplified using 'universal' primers and then sequenced. The universal primers were carefully designed to hybridize to highly conserved sites in order to allow the amplification of mtDNA across a broad range of species, yet sequence differentiation in a region sandwiched between the hybridization sites is sufficiently large to allow the discrimination of species.

The sequencing approach is very efficient because DNA samples from different species can be positively identified. However, confirmation of human samples is likely to be useful in routine forensic cases. Matsuda *et al.*, (2005) developed a PCR-based method in which DNA extracted from samples was amplified using 'human-specific' primers and then subjected to gel electrophoresis. The human-specific primers were designed to hybridize to human cyt *b* gene sites, which differed from those of the chimpanzee by 26% (Figure 8.3). The results of this method were determined simply by the presence (positive results) or the absence of a visible band (negative results), with no bands observed in DNA of animals, including non-human primates (chimpanzee, gorilla, Japanese monkey, crab-eating monkey) (Figure 8.4). Thus, samples producing a single band can be reasonably interpreted as being of human origin. Samples producing no visible bands, however, are inconclusive. In such cases, the employment of other cyt *b* gene primers as a positive control, and if necessary, the subsequent sequence analysis may achieve conclusive results (Parson *et al.*, 2000).

Single nucleotide polymorphism (SNP) typing

A random match probability of mtDNA types (HV-I/HV-II types) is estimated to be about 0.5–1% (Lutz *et al.*, 1998; Budowle *et al.*, 1999; Umetsu and Yuasa, 2005). However, this value is an average of all mtDNA types, and for a particular mtDNA type the chance that a random individual will share the same type depends greatly upon the relative rarity of that particular mtDNA type (Holland and Parsons, 1999). Indeed, the overall distribution of mtDNA types in many populations studied to date is highly skewed towards very rare types (Parsons and Coble, 2001). For example, of 1175 different mtDNA types in the US Caucasian populations, 982 types are unique (Parsons and

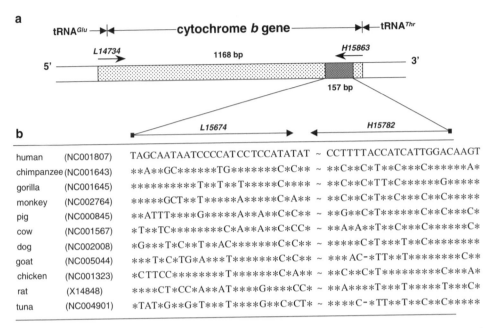

Figure 8.3 Schematic representation of the human mitochondrial cyt *b* gene and the amplification strategy for human identification. (a) Positions and orientations of the primers are indicated by arrows; (▨) the entire cyt *b* gene amplified by the *L14734/H15863* primer pair; (▨) the regions targeted for human-specific amplification using the *L15674/H15782* primer pair. (b) Comparison of human and various animal nucleotide sequences of regions targeted by primers *L15674* and *H15782*. GenBank accession numbers are given in parentheses. Reprinted from Matsuda *et al.* (2005) with permission from Elsevier

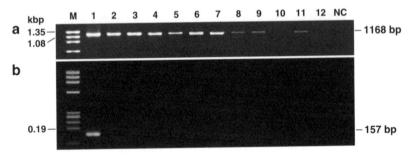

Figure 8.4 Amplification products of the primary and nested PCR. Amplified products were electrophoresed on 3% agarose gel and stained with ethidium bromide. (a) Primary PCR with outer primers (*L14734/H15863*): M, HaeIII digest marker (TaKaRa); lane 1, amplified product from human DNA; lane 2, chimpanzee; lane 3, gorilla; lane 4, Japanese monkey; lane 5, crab-eating monkey; lane 6, pig; lane 7, cow; lane 8, dog; lane 9, goat; lane 10, chicken; lane 11, rat; lane 12, tuna; NC, negative control (no template DNA was added). (b) Nested PCR with inner primers (*L15674/H15782*): PCR products were amplified from the products in the corresponding lanes of (a). Reprinted from Matsuda *et al.* (2005) with permission from Elsevier

Coble, 2001). When these very rare types are observed in both evidence and reference samples, it increases the likelihood that the two samples are from the same source. The increased likelihood of a positive match may often be sufficient to make a definite identification, especially if taken in conjunction with additional physical and/or circumstantial evidence. However, there are some relatively common mtDNA types. In the same population, the frequency of the most common mtDNA type, '263G, 315.1C' is 7%, and there are 13 additional mtDNA types with frequencies of 0.5% or larger (Parsons and Coble, 2001). If both the evidence and reference samples contain the same relatively common type of mtDNA, it does not increase the likelihood of the two samples having come from the same source. Thus, we must seek further biological evidence that will allow us to make a more definite conclusion regarding whether the two samples match. While the probabilities at individual STRs can be multiplied to obtain the total likelihood of a match, the HV-I and HV-II loci must be treated as a single locus, thereby reducing the power of obtaining a highly significant match.

To increase individual discrimination power, polymorphisms residing outside two hypervariable regions (HV-I and HV-II) have been increasingly explored. One approach, which has been promoted by SWGAM (the Scientific Working Group on DNA Analysis Methods) (Allard *et al.*, 2005) and by Carracedo and his colleagues (Quintáns *et al.*, 2004), is to expand targeting regions from the 600 bp HV-I and HV-II region to the entire 1100 bp control region. Using this approach, over 200 SNP sites were found, and, based upon the SNP profiles, phylogenically related mtDNA were grouped into haplogroups. The term 'SNP' was initially introduced to indicate a single DNA base substitution that is observed with a frequency of at least 1% in a given population. Currently, however, this term refers to any SNP, including insertions and deletions, and the 1% frequency prerequisite has been eliminated. It is noted that, with the exception of length polymorphisms (C-stretch), HV-I and HV-II polymorphisms are regarded as a collection of densely clustered SNPs.

Another attempt to increase individual discrimination power is to profile SNPs across the entire mtDNA. Even though the density of SNPs in the coding region is low due to functional restraints, the coding region is 14 times larger than the control region, therefore the number of SNPs in the coding region is comparable to the number of SNPs found in the control region. However, it is impossible to sequence the entire mtDNA, therefore the accumulation of data regarding SNP positions over the entire mtDNA (Parsons and Coble, 2001; Hall *et al.*, 2005) and the selection of an efficient method to spot the targeted SNPs (Sobrino *et al.*, 2005) will increase individual discrimination power considerably. Among various methods applicable to such a task, the authors feel that microarray (Divne and Allen, 2005) is the most promising in that it allows fast simultaneous detection of a large number of SNPs.

8.5 References

Allard, M.W., Polanskey, D., Miller, K., Wilson, M.R., Monson, K.L. and Budowle, B. (2005) Characterization of human control region sequences of the African American SWGDAM forensic mtDNA data set. *Forensic Sci. Int.* **148**: 169–179.

Anderson, S., Bankier, A.T., Barrell, G., de Bruijn, M.H.L., Coulson, A.R., Drouin, J., Eperon, I.C., Nierlich, D.P., Roe, B.A., Sanger, F., Schreier, P.H., Smith, A.J.H., Staden, R. and Young, I.G. (1981) Sequence and organization of the human mitochondrial genome. *Nature* **290**: 457–465.

Andrews, R.M., Kubacka, I., Chinnery, P.F., Lightowlers, R.N., Turnbull, D.M. and Howell, N. (1999) Reanalysis and revision of the Cambridge reference sequence for human mitochondrial DNA. *Nat. Genet.* **23**: 147.

Brown, T.A., Cecconi, C., Tkachuk, A.N., Bustamante, C. and Clayton, D.A. (2006) Replication of mitochondrial DNA occurs by strand displacement with alternative light-strand origins, not via a strand-coupled mechanism. *Genes Dev.* **19**: 2466–2476.

Budowle, B., Allard, M.W., Wilson, M.R. and Chakraborty, R. (2003) Forensic and mitochondrial DNA: application, debates, and foundations. *Annu. Rev. Genom. Hum. Genet.* **4**: 119–141.

Budowle, B., Wilson, M.R., DiZinno, J.A., Stauffer, C., Fasano, M.A., Holland, M.M. and Monson, K.L. (1999) Mitochondrial DNA regions HVI and HVII population data. *Forensic Sci. Int.* **103**: 23–35.

Butler, J.M. and Levin, B.C. (1998) Forensic applications of mitochondrial DNA. *Trends Biotechnol.* **16**: 158–162.

Carracedo, A., Bär, W., Lincoln, P., Mayr, W., Morling, N., Olaisen, B., Schneider, P., Budowle, B., Brinkmann, B., Gill, P., Holland, M., Tully, G. and Wilson, M. (2000) DNA Commission of the International Society for Forensic Genetics: guidelines for mitochondrial DNA typing. *Forensic Sci. Int.* **110**: 79–85.

Divne, A.M. and Allen, M. (2005) A DNA microarray system for forensic SNP analysis. *Forensic Sci. Int.* **154**: 111–121.

Embley, T.M. and Martin, W. (2006) Eukaryotic evolution, changes and challenges. *Nature* **440**: 623–630.

Giles, R.E., Blanc, H., Cann, H.M. and Wallace, D.C. (1980) Maternal inheritance of human mitochondrial DNA. *Proc. Natl. Acad. Sci. USA* **77**: 6715–6719.

Gray, M.W., Burger, G. and Lang, B.F. (1999) Mitochondrial evolution. *Science* **283**: 1476–1481.

Hall, T.A., Budowle, B., Jiang, Y., Blyn, L., Eshoo, M., Lowery, K.A.S., Sampath, R., Drader, J.J., Hannis, J.C., Harrell, P., Samant, V., White, N., Ecker, D.J. and Hofstadler, S.A. (2005) Base composition analysis of human mitochondrial DNA using electrospray ionization mass spectrometry: a novel tool for the identification and differentiation of humans. *Anal. Biochem.* **344**: 53–69.

Hauswirth, W.W., Van de Walle, M.J., Laipis, P.J. and Olivo, P.D. (1984) Heterogeneous mitochondrial DNA D-loop sequences in bovine tissue. *Cell* **37**: 1001–1007.

Hayashi, J., Takemitsu, M., Goto, Y. and Nonaka, I. (1994) Human mitochondria and mitochondrial genome function as a single dynamic cellular unit. *J. Cell Biol.* **125**: 43–50.

Hershko, A. and Ciechanover, A. (1998) The ubiquitin system. *Annu. Rev. Biochem.*, **67**: 425–479.

Holt, I.J., Harding, A.E., Petty, R.K.H. and Morgan-Hughes, J.A. (1990) A new mito-chondrial disease associated with mitochondrial DNA heteroplasmy. *Am. J. Hum. Genet.* **46**: 428–433.

Holland, M.M. and Parsons, T.J. (1999) Mitochondrial DNA sequence analysis – validation and use for forensic casework. *Forensic Sci. Rev.* **11**: 21–50.

Kaneda, H., Hayashi, J., Takahama, S., Taya, C., Lindahl, K.F. and Yonekawa, H. (1995) Elimination of paternal mitochondrial DNA in intraspecific crosses during early mouse embryogenesis. *Proc. Natl. Acad. Sci. USA* **92**: 4542–4546.

Kocher, T.D., Thomas, W.K., Meyer, A., Edwards, S.V., Pääbo, S., Villablanca, F.X. and Wilson, A.C. (1989) Dynamics of mitochondrial DNA evolution in animals: amplification and sequencing with conserved primers. *Proc. Natl. Acad. Sci. USA* **86**: 6196–6200.

Lutz, S., Weisser, H.J., Heizmann, J. and Pollak, S. (1998) Location and frequency of polymorphic positions in the mtDNA control region of individuals from Germany. *Int. J. Legal Med.* **111**: 67–77.

Matsuda, H., Seo, Y., Kakizaki, E., Kozawa, S., Muraoka, E. and Yukawa, N. (2005) Identification of DNA of human origin based on amplification of human-specific mitochondrial cytochrome *b* region. *Forensic Sci. Int.* **152**: 109–114.

Meyer, S., Weiss, G. and Haeseler, A. (1999) Pattern of nucleotide substitution and rate heterogeneity in the hypervariable regions I and II of human mtDNA. *Genetics* **152**: 1103–1110.

Michikawa, Y., Mazzucchelli, F., Bresolin, N., Scarlato, G. and Attardi, G. (1999) Aging-dependent large accumulation of point mutations in the human mtDNA control region for replication. *Science* **286**: 774–779.

Moretti, T.R., Baumstark, A.L., Defenbaugh, D.A., Keys, K.M., Smerick, J.B. and Budowle, B. (2001) Validation of short tandem repeats (STRs) for forensic usage: performance testing of fluorescent multiplex STR systems and analysis of authentic and simulated forensic samples. *J. Forensic Sci.* **46**: 647–660.

Nishimura, Y., Yoshinari, T., Naruse, K., Yamada, T., Sumi, K., Mitani, H., Higashi-yama, T. and Kuroiwa, T. (2006) Active digestion of sperm mitochondrial DNA in single living sperm revealed by optical tweezers. *Proc. Natl. Acad. Sci. USA* **103**: 1382–1387.

Parson, W., Pegoraro, K., Niederstätter, H., Föger, M. and Steinlechner, M. (2000) Species identification by means of the cytochrome *b* gene. *Int. J. Legal Med.* **114**: 23–28.

Parsons, T.J. and Coble, M.D. (2001) Increasing the forensic discrimination of mito-chondrial DNA testing through analysis of the entire mitochondrial DNA genome. *Croat. Med. J.* **42**: 304–309.

Pinz, K.G. and Bogenhagen, D.F. (1998) Efficient repair of abasic sites in DNA by mitochondrial enzymes. *Mol. Cell. Biol.* **18**: 1257–1265.

Quintáns, B., Iglesias, Á., Salas, A., Phillips, C., Lareu, M.V. and Carracedo, A. (2004) Typing of mitochondrial DNA coding region SNPs of forensic and anthropological interest using SnaPshot minisequencing. *Forensic Sci. Int.* **140**: 251–257.

Sato, A., Nakada, K., Akimoto, M., Ishikawa, K., Ono, T., Shitara, H., Yonekawa, H. and Hayashi, J. (2005) Rare creation of recombinant mtDNA haplotypes in mammalian tissues. *Proc. Natl. Acad. Sci. USA* **102**: 6057–6062.

Shuster, R.C., Rubenstein, A.J. and Wallace, D.C. (1988) Mitochondrial DNA in anucle-ate human blood cells. *Biochem. Biophys. Res. Commun.* **155**: 1360–1365.

Sobrino, B., Brión, M. and Carracedo, A. (2005) SNPs in forensic genetics: a review on SNP typing methodologies. *Forensic Sci. Int.* **154**: 181–194.

Stewart, J.E., Fisher, C.L., Aagaard, P.J., Wilson, M.R., Isenberg, A.R., Polanskey, D., Pokorak, E., DiZinno, J.A. and Budowle, B. (2001) Length variation in HV2 of the human mitochondrial DNA control region. *J. Forensic Sci.* **46**: 862–870.

Stoneking, M. (1994) Mitochondrial DNA and human evolution. *J. Bioenerg. Biomem.* **26**: 251–259.

Stoneking, M. (2000) Hypervariable sites in the mtDNA control region are mutational hotspots. *Am. J. Hum. Genet.* **67**: 1029–1032.

Stoneking, M., Hedgecock, D., Higuchi, R.G., Vigilant, L. and Erlich, H.A. (1991) Population variation of human mtDNA control region sequences detected by enzymatic amplification and sequence-specific oligonucleotide probes. *Am. J. Hum. Genet.* **48**: 370–382.

Taanman, J.W. (1999) The mitochondrial genome: structure, transcription, translation and replication. *Biochim. Biophys. Acta* **1410**: 103–123.

Tully, L.A. and Levin, B.C. (2000) Human mitochondrial genetics. *Biotechnol. Genet. Eng. Rev.* **17**: 147–177.

Umetsu, K. and Yuasa, I. (2005) Recent progress in mitochondrial DNA analysis. *Legal Med.* **7**: 259–262.

Wilson, M.R., DiZinno, J.A., Polanskey, D., Replogle, J. and Budowle, B. (1995) Validation of mitochondrial DNA sequencing for forensic casework analysis. *Int. J. Legal Med.* **108**: 68–74.

9
Y-Chromosomal markers in forensic genetics

Manfred Kayser

9.1   Introduction

Analysis of the human Y chromosome in forensics has three main applications: male sex identification, male lineage identification and identification of the geographical origin of male lineages. A male individual is identified, based on DNA evidence, by detecting male-specific parts of the Y chromosome in crime scene samples. Male lineage identification (a male lineage is defined as a male individual together with all his paternal male relatives) is performed by typing male-specific Y-chromosomal DNA polymorphisms in the crime scene samples and searching for matching profiles in suspects. The geographical origin or, in other words, genetic ancestry of male lineages is revealed by using Y-chromosomal markers with specific geographical distributions, as determined from reference databases. The value of Y-chromosomal markers for male identification in forensics is underlined by the fact that the vast majority of violent crimes are committed by males and that almost all cases of sexual assault involve males as perpetrators. Consequently, Y-chromosomal markers are increasingly being used by forensic laboratories. This is reflected in the large increase in the number of publications dealing with Y-chromosomal markers in the forensic science literature over recent years (Figure 9.1). A number of milestone discoveries in human genetics made it possible to use human Y-chromosomal markers for forensic applications. In this chapter I will describe the applications of Y-chromosomal markers to modern forensics, together with some of the key discoveries in human molecular and population genetics that have allowed such applications, and finally I will give a brief outlook on the future of Y-chromosomal markers in forensics.

Molecular Forensics.   Edited by Ralph Rapley and David Whitehouse
Copyright 2007 by John Wiley & Sons, Ltd.

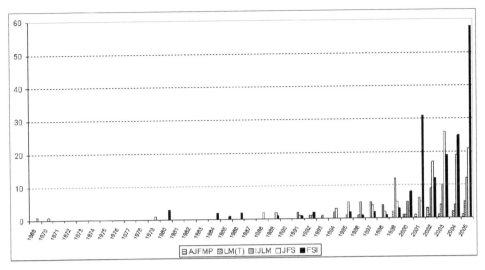

Figure 9.1 Number of publications per year dealing with Y-chromosome markers in five scientific forensic journals as revealed by the PubMed database (*http://www.ncbi.nlm.nih.gov/entrez/query.fcgi?DB=pubmed*) and selecting from key word queries: 'Y-chromosome, Y-STR, Y STR and sex': *International Journal of Legal Medicine* (IJLM) PubMed record since 1991, *Forensic Science International* (FSI) PubMed since 1978, *Journal of Forensic Sciences* (JFS) PubMed since 1965, *Legal Medicine* (LM) PubMed since 1999 and *American Journal of Forensic Medicine and Pathology* (AJFMP) PubMed since 1980

9.2   Identification of the male sex

Up until the beginning of the last century it was generally believed that in humans sex was determined by environmental factors such as maternal nutrition. With the discovery of the human X and Y chromosomes in the early 1920s (Painter, 1923) it was first assumed that the number of X chromosomes determined the sex of a human individual. It took an additional 36 years to establish that sex determination in humans and other mammals is independent from the number of X chromosomes and that the presence of the Y chromosome is responsible for the male sex (Jacobs and Strong, 1959). During more than three decades of continued, intensive research into the molecular basis of human male sex determination, a series of putative candidate sequences on the human Y chromosome have been established, e.g. simple repetitive sequences (Epplen *et al.*, 1983; Kiel-Metzger *et al.*, 1985). Finally in the early 1990s the so-called 'sex determining region Y' or SRY gene (Sinclair *et al.*, 1990) was identified and it was subsequently shown that the transfer and expression of SRY in female mouse embryos led to the development of testicles (Koopman *et al.*, 1991). Today it is generally accepted that the SRY gene expresses the testis-determining factor and is the key gene responsible for male sex formation in humans and other mammals (Berta *et al.*, 1990).

These (and other) discoveries in human genetics opened the door for forensic DNA-based human sex identification. In the 1970s luminescence microscopy was used for detection of the human Y chromosome in forensic material (Pearson and Bobrow, 1970; Radam and Strauch, 1973). Later, advances in molecular genetics allowed more sensitive detection of the Y chromosome using, for example, Y alphoid DNA (Stalvey and Erickson, 1987). The breakthrough for DNA-based sex identification in forensics came with the introduction of the polymerase chain reaction (PCR) for the sensitive detection of various regions on the human Y chromosome (Witt and Erickson, 1989), including the amelo-genin gene (Akane *et al.*, 1991).

Although the detection of DNA from the non-recombining region of the Y chromosome identifies the presence of male material, not detecting Y-specific DNA does not mean that a sample contains only female material. This is because a negative result in a Y chromosome DNA test can have other reasons than there being no Y chromosome present in the sample being investigated, e.g. technical failures, no amplifiable DNA, etc. For this reason, combined tests to detect both Y-chromosomal and X-chromosomal DNA were developed. Of these, the amelogenin gene test has become the most established in forensic laboratories (Akane *et al.*, 1991). The amelogenin gene is present on both the X and the Y chromosome and the test is based on a length polymorphism within the gene itself differentiating the Y-chromosomal from the X-chromosomal copy. Nowadays this test is included in many commercial kits for the DNA-based identification of human individuals. However, it should be noted that the reliability of the test has been criticized due to the occurrence of Y-chromosomal deletions that can include the amelogenin gene (Santos *et al.*, 1998). Although the frequency of such deletions is generally low, their incidence can be increased in certain populations due to events in the population history (Santos *et al.*, 1998; Steinlechner *et al.*, 2002; Thangaraj *et al.*, 2002). Furthermore, due to the highly repetitive molecular structure of the Y chromosome, deletions are known from many Y-chromosomal regions. Therefore, the reliability of DNA-based sex tests can be improved by increasing the number of Y loci tested (Santos *et al.*, 1998). Naturally, the detection of Y-chromosomal DNA poly-morphisms as used to identify male lineages because of their property to carry genetic variation between male lineages (see next chapter) is informative for male sex identification.

9.3 Identification of male lineages

Most of the human Y chromosome (i.e. the non-recombining region of the Y chromosome, NRY) is male specific and is inherited unchanged from fathers to sons, unless a rare mutational event occurs. DNA recombination, a genetic process that reshuffles genetic material between homologue chromosomes to create additional variation, is not acting on the NRY because of the absence of a homologue chromosome. Due to the lack of recombination, a Y-chromosomal mutation creating a new allele is always inherited by male offspring in subsequent generations. On the other hand, the lack of recombination also means that all male relatives carry the same Y chromosome, independent of the degree of paternal relationship. This makes Y chromosome polymorphisms very useful for male identification but also means that male lineages (i.e. groups of paternally related males), but not individual males, can be identified, at least with the currently available Y markers.

The use of Y-chromosomal polymorphisms for male identification in forensics started 40 years ago with the analysis of whole Y-chromosome length polymorphisms to detect exclusion constellations in paternity cases (Nuzzo *et al.*, 1966), even though the molecular basis of the underlying polymorphisms was unknown at the time. Later, and with increased knowledge about the molecular biology of the human Y chromosome, the use of whole Y-chromosome length differences was abolished from forensic applications due to the discovery of Y-chromosome length differences between cells from the same individual (Daiger and Chakraborty, 1985). The real breakthrough came in the early 1990s with the identification of the first Y-chromosomal microsatellite or short tandem repeat (Y-STR) polymorphism, DYS19 (Roewer *et al.*, 1992), and its immediate application to a rape case, revealing an exclusion constellation (Roewer and Epplen, 1992). However, as known from the application of autosomal STRs to forensics, the value of a single marker for human individualization is limited. Instead, many STRs are needed to achieve high resolution and confidence. In 1997 a first attempt towards the forensic application of Y-STR haplotypes was made by the Forensic Y-Chromosome Research Group coordinated by Lutz Roewer from the Humboldt University in Berlin with the characterization of 13 Y-chromosomal STRs in 3825 unrelated males from 48 population samples (Kayser *et al.*, 1997). This study was recently rated as the second most highly cited publication in the five leading forensic science and legal medicine journals, according to the ISI Web of Science database (Jones, 2007), but in fact it is the most cited paper ever published in a leading forensic journal (306 citations) based on a database query in July 2006, reflecting the success of Y-STR markers, especially in forensic science.

By 1997, 13 Y-chromosomal STRs were available for forensic applications. Of these, nine describe the so-called 'minimal haplotype' recommended by the International Forensic Y User Group as the minimal set of Y-STRs to be used for human male individualization in forensics (Kayser *et al.*, 1997). The advantage of genetic markers from the non-recombining part of the Y chromosome compared with those from any other chromosome is that single marker information can be combined as haplotype information since the male-specific part of the Y chromosome is inherited completely linked from fathers to sons. For forensic applications this means that the multiplication of single locus allele frequencies for obtaining combined DNA profile matching probabilities cannot be applied to Y-chromosomal markers, but instead compound haplotype frequencies must be used for establishing matching probabilities in cases of non-exclusions. The combination of single loci in compound Y-chromosomal haplotypes, such as using the nine Y-STRs from the minimal haplotype, leads to an enormous increase in informativity. Consequently, the number of individuals that need to be investigated for obtaining representative haplotype frequencies is expected to be enormous and much larger than needed for autosomal markers. This has led to the establishment of Y-STR haplotype databases for obtaining more accurate and reliable haplotype frequencies. The largest database is the 'Y-Chromosome Haplotype Reference Database, YHRD'. This database started out as a European initiative (Roewer *et al.*, 2001), was later expanded by mirror databases for U.S. populations (Kayser *et al.*, 2002), and Asian populations (Lessig *et al.*, 2003) and today exists as a combined and further expanded database with 40 108 haplotypes in a set of 320 populations worldwide, including population samples from all continental regions (Release '19' from August 2006).

Numerous laboratories, mostly from the international forensic genetics community, have contributed Y-STR haplotype data under controlled quality criteria to the YHRD and the number of contributed haplotypes is constantly increasing. The YHRD allows complete and partial Y-STR haplotype profiles to be searched for population-based and region-based frequencies, and provides useful graphical representations of the geographical distribution of the respective haplotypes, as well as lists with the number of matches per population sample. In addition to the search function, information about typing methods, molecular characteristics, including mutation rate estimates, and several statistical tools, including a haplotype frequency surveying method, are available through the public-domain website of the database (*http://www.yhrd.org*). This makes the YHRD unique not only in size but also in data authenticity, compared with databases established from published Y-STR data or other databases without quality control requirements. Due to the introduction of various commercial Y-STR kits (Yfiler from Applied Biosystems, Power-Plex Y from Promega, genRES DYSplex from Serac, Y-PLEX from Reliagene, Menplex Argus Y from Biotype), company-based Y-STR haplotype databases have become available recently, mostly collecting data from individuals living in the USA:

http://www.promega.com/techserv/tools/pplexy/
http://www.appliedbiosystems.com/yfilerdatabase/
http://www.reliagene.com/index.asp?menu_id=rd&content_id=y_frq

In the future it would be desirable if all data could be included in a single database allowing user-friendly single access for a comprehensive Y-STR haplotype frequency search.

The number of scientifically known Y-STRs has dramatically increased over recent years. In 2004, results from a comprehensive survey of Y-STRs were published using the nearly complete Y chromosome sequence for a systematic search for all useful Y-STR markers (Kayser *et al.*, 2004). In this study, 166 previously unknown Y-STR markers were found, increasing the total number of verified Y-STRs to 215 (Figure 9.2). Although the 9–11 commonly used Y-STRs provide high haplotype diversity, and thus high probability of male lineage identification, additional Y-STRs will increase the haplotype discrimination, depending on the marker added and the population analysed (Beleza *et al.*, 2003; Park *et al.*, 2005; Turrina *et al.*, 2006). Furthermore, studies have reported population samples with a high number of identical 9–16 loci Y-STR haplotypes as a result of severe bottlenecks in the history of those populations, e.g. 14% of males in a Pakistani population sample (Mohyuddin *et al.*, 2001) and 13% of males in a Finish population sample (Hedman *et al.*, 2004) share the same 16-loci Y-STR haplotype. Therefore, in cases of matching haplotypes, typing additional Y-STRs can be useful for forensic applications, and currently many of the newly described Y-STRs are investigated with respect to their haplotype discrimination potential, population genetic diversity, mutation rates, as well as their suitability for multiplex analyses.

A number of Y-STR markers are located in multicopy regions of the Y chromosome and thus consistently show more than one male-specific allele (Kayser *et al.*, 2004). These multicopy Y-STRs, although often very variable due to the simultaneous detection of multiple polymorphic loci (Redd *et al.*, 2002), are less

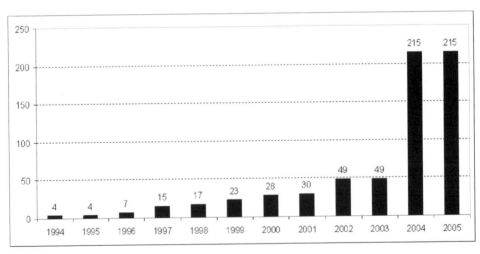

Figure 9.2 Cumulative number of Y-STRs, as published in scientific journals. Markers are counted separately only when amplifiable separately; consequently, multicopy Y-STRs that cannot be typed separately are counted only once. Those Y-STR markers that were submitted to public databases but not verified through scientific publications were not included

useful for forensic stain analyses since they might cause interpretation difficulties in mixed stains with more than one male involved (Butler *et al.*, 2005). Some of these multicopy markers are located in Y-chromosomal regions where the number of copies is assumed to be associated with male fertility problems (such as AZF). Consequently, a Y-STR profile including these loci can potentially be informative for the fertility status of a man (Bosch and Jobling, 2003). Such loci should be omitted from forensic tests because of the additional information they potentially reveal.

The most important application of Y-chromosome markers for male lineage identification in forensics is in cases of sexual assault. The nature of the material available from sexual assault cases, usually mixed stains from the female victim's epithelial cells and the male perpetrator's sperm cells, makes autosomal STR profiling challenging. Normally, to separate the male and female genetic components, differential lysis is applied to extraction DNA from mixed stains. Often, and especially when the number of sperm cells is low, this approach fails, resulting in potential overlap of the victim's and perpetrator's autosomal STR profiles, making male perpetrator identification impossible. Recent technological advances, e.g. using laser dissection microscopy to specifically collect sperm cells, are promising but the success of autosomal STR analysis from such material depends on the number of sperm cells collectable in each particular case. If the number of sperm cells is low, technical problems of low copy number (LCN) analysis are expected, and the approach will not be successful when no sperm cells can be collected. However, the specific detection of male DNA by analysing Y-chromosomal STRs in principle avoids the problem of profile overlap (since females do not carry a Y chromosome) and additionally is much more sensitive. Mixing experiments have shown that Y-STRs can still be amplified successfully and reliably up to male–female DNA mixtures of 1:2000 (Prinz *et al.*, 1997). In addition, even in the absence of sperm (i.e. in cases of oligospermic or azoospermic males involved) but the presence of mostly or only male epithelial cells in a mixed stain, Y-STR analysis has proven to be highly successful (Betz *et al.*, 2001). It has also been shown that Y-STR haplotype profiling in rape cases can be successful from cervicovaginal samples recovered up to 4 days post-coitus (Hall and Ballantyne, 2003).

The use of Y-STR markers in forensics is regulated by two recommendations of the DNA Commission of the International Society of Forensic Genetics in collaboration with expert Y chromosome scientists (Gill *et al.*, 2001; Gusmao *et al.*, 2006). Because of the above-mentioned advantages and the enormous research effort that has been undertaken, Y chromosome markers (especially Y-STRs) are routinely used for male lineage identification in many forensic laboratories all over the world, and have been for many years already. One case shall be mentioned as an example to demonstrate the power of Y-STRs in forensic male lineage identification. To identify a serial rapist who had raped 14 young women and murdered one of them in north-western Poland, Y-STR haplotype profiles were obtained from >400 suspects as part of an elimination process (Dettlaff-Kakol and Pawlowski, 2002). A man was identified with a

Y-STR haplotype profile identical to that obtained from the crime scene, but with an autosomal STR profile matching that of the crime scene in only 9 of 10 markers. This finding suggested that the perpetrator must be one of this man's close relatives and DNA analysis of his brother revealed a complete match of both the autosomal STR and Y-STR profiles, identifying the brother as the perpetrator. It should be pointed out that Y chromosome (or mtDNA)-based mass screenings (which are on a voluntary basis) are also seen critically since legislation in many countries provides the right to refuse testimony in cases where close relatives are involved. However, Y chromosome (and mtDNA)-based male (and female) lineage identification is able to identify close and distant male (and female) relatives.

Although there is general agreement on the use of Y-STR markers to exclude suspects, there is still an ongoing discussion about how to use Y-STR haplotype information in the courtroom when a match between the crime scene sample and a suspect is established. Usually it is common practise to place some significance on the probability of such a match. This can be achieved by a method that extrapolates frequency estimates based on observed data stored in a database such as the YHRD (Krawczak, 2001; Roewer *et al.*, 2000). This 'haplotype surveying method', available via the YHRD website, generates estimates of the prior and posterior frequency distributions of Y-STR haplotypes (Roewer *et al.*, 2000). A simplified use of Y-STR haplotype databases for obtaining confidence in the statistical meaning of a haplotype match can be obtained from mismatch distribution analysis (Pereira *et al.*, 2002), whereas a more conservative approach is to simply count the number of times the haplotype exists in the database and establish a confidence interval by taking into account the size of the database (Budowle *et al.*, 2003). However, some scientists argue that existing Y-STR databases are not representative of real populations because of their limited size and because the databases are normally based on unrelated individuals, whereas real populations are not only bigger in size but also do contain related individuals. The latter is especially important for Y chromosome markers since close and distantly related men can share the same Y-STR haplotype. Such a perspective leads to the most conservative use of Y-STR haplotype information in cases of established matches, stating that the suspect cannot be excluded from being the donor of the crime scene sample (de Knijff, 2003). It is noteworthy that official recommendations by the DNA Commission of the International Society of Forensic Genetics on the estimation of the weight of the evidence of Y-STR typing have not yet been provided but are expected in the near future, as announced elsewhere (Gusmao *et al.*, 2006).

9.4 Identification of a male's paternity

The second forensic application offered by Y chromosome markers is in testing for the paternity of male offspring. Already in 1985, and therefore seven years before the first hypervariable Y-STR marker was discovered, it was suggested based on statistical considerations that the use of polymorphic Y markers should increase the chance to detect non-paternity compared with their autosomal counterparts (Chakraborty, 1985). Polymorphic Y-chromosomal markers are especially useful in deficiency cases where the alleged father of a male child is deceased. Such cases can only be solved via autosomal STR profiling with a high degree of certainty if both parents of the putative father are available for DNA analysis. Only the complete genotypes of the grandparents allow a reliable reconstruction of the STR alleles inherited from the deceased putative father to his child, given the mother's profile. However, many cases are brought to paternity testing where none or only one parent of the deceased alleged father is available for DNA analysis. In those cases involving a male child, Y-STR haplotyping can identify the biological father if a biological male relative of any grade of relationship is available for testing to replace the deceased alleged father in the Y chromosome DNA analysis. Since Y-chromosomal STR haplotypes are identical between male relatives (unless rare mutation events occur), finding matching haplotypes between the male child and any biological paternal relative of the putative father will provide evidence in favour of the biological paternity of the deceased putative father (or any of his contemporary male relatives). Conversely, finding haplotype differences between the child and the male relative will exclude the deceased alleged father from paternity. However, the number of generations that separate the alleged father from his male relative used for Y chromosome analysis will increase the probability that mutations will introduce differences between the male relative and the child, despite the deceased male being the true father of the child. Thus, mutation rate estimates of the Y-STRs used for testing need to be taken into account in calculating paternity probabilities (Rolf et al., 2001). Therefore, knowledge about mutation rates for the Y-STRs used in forensics is important (Kayser and Sajantila, 2001).

The first study to estimate mutation rates for Y-STRs used in forensics applied deep-rooting pedigrees and revealed an average rate of 2.1×10^{-3} mutations per locus per generation (Heyer et al., 1997). In principle, mutation rate estimates should include the uncertainty caused by potential non-biological paternity, being especially crucial in pedigree studies where paternity cannot be established directly. However, in this study the biological paternal relationships within the pedigrees used for Y-STR mutation rate estimation were confirmed by additional analysis of the Y-chromosomal minisatellite MSY1 (Jobling et al., 1999). The first comprehensive study establishing Y-STR mutation rates based on father–son pairs of (autosomal) DNA-proven biological paternity revealed an average rate of 2.8×10^{-3} mutations per locus per generation (Kayser et al., 2000b). Subsequent, additional studies using father–son pairs confirmed previously obtained mutation rates (Ballard et al., 2005; Budowle et al., 2005; Dupuy et al., 2004; Gusmao et al., 2005) and a summary of mutation rate estimates for Y-STR markers commonly used in forensics is available from the YHRD website. Given the hypervariability of Y-STR haplotypes with sufficient loci included (e.g. the minimal haplotype), paternally unrelated males normally show differences at many Y-STRs. If two minimal haplotypes show differences at only one or two loci involving only one or two repeats, a relationship between the two respective male individuals needs to be considered, given the available knowledge about Y-STR mutation rates. It has been suggested that exclusion constellations at three or more Y-STRs need to be established before an exclusion of paternity can be concluded (Kayser *et al.*, 1998; Kayser and Sajantila, 2001). Mutations at STRs, independent of whether they are located on the autosomes or on the sex chromosomes, are results of errors during DNA replication and mismatch repair. Single-strand slippage within the repetitive sequence of the STR locus can lead to a gain or a loss of repeats, usually of single repeat units.

Another feature of Y-STRs that is of relevance for forensic applications (although not for paternity testing) concerns the rare occurrence of additional Y-STR alleles (Kayser and Sajantila, 2001). Two mutational events lead to the observation of additional Y-STR alleles: first, a Y-chromosomal duplication including the STR locus occurs, and subsequently a slippage mutation results in allelic differences between the original STR and the copy. Additional Y-STR alleles were observed at almost all Y-STRs used in forensics (Butler *et al.*, 2005) and can be misinterpreted as the involvement of multiple males when detected in a crime scene sample.

In principle, the absence of recombination between Y-specific markers allows the identification of non-biological paternity after many (male) generations in pedigree analysis. Thus, disputed paternity cases from historical times involving male offspring can be resolved today by testing Y-chromosomal polymorphisms in true paternal male descendants. This has been done in many cases, with the most prominent being that of the former U.S. President Thomas Jefferson and the children of Sally Hemmings, one of his slaves (Foster *et al.*, 1998). It could be shown that Y-chromosomal profiles based on Y-SNPs, Y-STRs and a Y minisatellite of a fourth-generation male descendant of Easton Hemings Jefferson, one of Sally Hemmings sons, and four sixth- and seventh-generation descendants of Field Jefferson, Thomas Jefferson's father's brother, were identical. From the Y-chromosome data it was concluded that either Thomas Jefferson or one of his contempory male-line relatives, including his brother Randolph, had fathered Easton Hemings Jefferson. Unfortunately no living male descendants of Randolph Jefferson were available for testing.

In the same way that Y chromosome DNA analysis can be used for paternity testing and forensic stain analysis, it can also be used for identifying the biological remains of missing persons, including cases of mass disasters (Corach *et al.*, 2001). Reference material of known living relatives is needed, as for autosomal DNA analysis, but the advantage of Y (and mitochondrial DNA) markers over autosomal markers is that relationships can be established and individuals identified, even if only reference samples of distant relatives are available for analysis. This can be highly relevant in mass disasters such as the 2004 Tsunami disaster in Southeast Asia, where entire families died and close relatives were therefore not available as references for autosomal DNA testing.

9.5   Identification of a male's geographical origin

The third forensic application of Y-chromosomal DNA polymorphisms is in the identification of the geographical origin or genetic ancestry of an unknown male individual. Since this is the most recent application of Y-chromosomal markers to forensics, it deserves a somewhat more detailed summary. Geographical origin or, in other words, genetic ancestry identification is important in forensic cases with no known suspects. In such cases it would be helpful for the police to be able to concentrate their investigation towards finding suspects from specific groups of individuals, i.e. people of a particular geographical origin (often the terms 'ethnic group' or 'ethnic identification' are used but are unfortunate since ethnicity is determined by more factors than geography). Genetic testing can, to a certain extent, provide such information, at least for some geographical regions of the world. However, in order to trace the suspect(s), the police would usually extrapolate information on particular externally visible characteristics from the DNA data providing information about the geographical origin. Assumptions about a suspect's looks based on his DNA-based geographical origin are strictly indirect and the entire approach is feasible only when a high correlation between a geographical region and an externally visible trait exists. For instance, there is a high correlation between human skin colour and latitude also leading to continental differentiation. As result, European geographical origin is usually strongly associated with light skin colour, whereas African genetic ancestry usually is with dark skin colour. Because of this strong association, it is somewhat justified to conclude a light skin colour appearance of a donor of a DNA sample when DNA typing reveals a European genetic origin, and a dark skin colour appearance from a DNA test revealing a African genetic origin. However, similarly strong correlations involving other phenotypic traits and geographical regions are rare.

DNA-based identification of geographical origins is usually performed by testing markers where a specific allele or haplotype is restricted to a certain geographical region or shows significant and large frequency differences between geographical regions. In general, frequency distributions of genetic markers arise when a mutation occurs in a single individual living in a particular geographical region and inherits the mutations to produced offspring who subsequently spread/migrate to other geographical regions or remain where they are. There can be several reasons why a marker increases in frequency so that it can be used for geographical origin identification. For instance, the mutation can have a beneficial effect on the individuals carrying it, resulting in reproductive success. Such effects of positive selection causing high marker frequencies are known, for instance, from genes responsible for or associated with resistance towards certain infectious diseases but can be expected from all genes with close environmental interactions and severe influence on survival and reproduction. In the case of resistance towards infectious diseases, the marker frequency depends on the strength of selection (benefit) but also on the frequency of the disease-causing organisms (e.g. mutations in genes expressed in red blood cells provide malaria resistance and are frequent in regions with a high incidence of malaria because of the high frequency of malaria-causing *Plasmodium* spp.). However, based on existing knowledge, positive selection is unlikely to have shaped Y chromosome diversity and the frequency distribution of Y markers basically depends on the mutation rate, the geographical region of occurrence of the mutation, cultural factors influencing the degree of male reproduction (i.e. residence and marriage patterns, warfare, etc.) and the migratory history of the respective (male) population.

With the availability of the first population data of Y-STRs it was noticed that – albeit rarely – some Y-STR marker alleles show a highly restricted geographical distribution, e.g. short DYS390 alleles in the Pacific region (Kayser *et al.*, 1997). Also, significant differences in the Y-STR haplotypes were found between geographically distant populations (Kayser *et al.*, 2001) as well as between geographically close populations, such as the Germans and the Dutch (Roewer *et al.*, 1996), although not between many other European groups (Roewer *et al.*, 2001). Within Europe at least three different groups of populations (called metapopulations in the YHRD) – Eastern Europeans, Western Europeans and Southeastern Europeans – were identified in the YHRD (Roewer *et al.*, 2005). Thus, based on the minimal Y-STR haplotype, information about which European region a male and his paternal ancestors originated from can be obtained, at least for some of the most characteristic haplotypes and those that show a more restricted distribution.

With the recent expansion of the YHRD to additionally include non-European population samples, continental information can also be obtained from Y-STR haplotype data, although such conclusions are still preliminary because of the low (but growing) number of non-European samples. There are a number of Y-STR haplotypes that show a continentally restricted frequency distribution. For instance, a YHRD search in August 2006 (Release '19') revealed that the Y-STR haplotype most characteristic for the Eastern European population cluster (DYS19, DYS389I, DYS389II, DYS390, DYS391, DYS392, DYS393, DYS385a–b: = 17, 13, 30, 25, 10, 11, 13, 10–14; Roewer *et al.*, 2005) is found in 192 out of 40 108 individuals in a set of 320 worldwide populations of which 186 (97%) are European (Plate 9.1a). From the remaining six matches, one (0.5%) is in Turkey, one (0.5%) in Kazakhstan and four (2%) in Hungarian Gypsies. This haplotype was not observed elsewhere in the world. Of the 186 European matches, 123 (66%) are found in the Eastern European metapopulation, as expected (most frequent all over Poland; somewhat frequent in Ukraine; rare in Lithuania, Latvia and Slovenia), 53 (28.5%) in the Western European metapopulation (mostly all over Germany; somewhat frequent in Czechia; rare in

Sweden and Italy), 7 (3.8%) in the South-Eastern European metapopulation (Greece, Hungary, Macedonia, Romania), 2 (1%) in US Americans and 1 (0.5%) from Argentina – the latter three men are of self-declared European descent.

The Y-STR haplotype most characteristic for the Western European meta-population (14, 13, 29, 24, 11, 13, 13, 11–14; Roewer *et al.*, 2005) was found in 820 out of 40 108 individuals in a set of 320 worldwide populations (Plate 9.1b), of which 731 (89%) are Europeans. From the 89 remaining matches 85 are of likely European ancestry through European admixture: 30 are from the USA, UK, Brazil or Columbia but from individuals of self-declared African ancestry, 3 are from African countries (Angola, Equatorial Guinea), 2 are from the UK but from individuals with self-declared Asian ancestry, 13 are from Reunion Creoles, 7 are from Ecuador Mestizos, 2 are from Ecuador Quichuas – all of these most likely indicate European Y-chromosomal admixture – and 28 are from U.S. Hispanics. Thus, altogether this haplotype is found in 99.5% of the matches in individuals with European ancestry; the remaining four matches are from Turkey, China, Georgia, and Hungarian Gypsy. This haplotype was not observed else-where in the world. As expected, this haplotype is most frequent in the Western Europeans with 466 (57%) matches (frequent all over Portugal, Spain, France, The Netherlands, Belgium, Ireland, UK, Germany, Switzerland, Italy; less fre-quent in Sweden, Denmark, Norway, U.S. Europeans; rare in Austria, Czechia, Estonia, Finland), and was also found with 16 matches (2%) in the Eastern European metapopulation (mostly Poland; rare in Slovenia and Germany), 16 matches in the Southeastern European metapopulation (2.3%) (rare in Greece, Hungary, Italy, Macedonia, Romania, Albania, Bulgaria) and 233 matches (28.4%) in Europeans from Argentina, Brazil, Colombia, South Africa and US Americans of European descent. In contrast, the Y-STR haplotype that is most frequent in US African Americans (15, 13, 31, 21, 10, 11, 13, 16, 17; Kayser *et al.*, 2002) was found with 41 matches in the YHRD (Plate 9.1c), of which 38 (92.6%) are Africans or men with known African descent: Cameroon, Bantu South Africa, Guinea, Mozambique, Egypt, African Americans from the USA, Brazil, Ecuador, Colombo and UK Afro-Caribbeans. The remaining 3 matches are from one Argentinean European, one US Hispanic and one Reunion Creole, most likely indicating African Y-chromosomal admixture.

Although, as can be seen, at least some Y-STR haplotypes are informative for geographical origin identification, the relatively high mutation rate of Y-STRs (Kayser *et al.*, 2000b) tends to randomize genetic ancestry signals over a large number of generations/long time span. Therefore it is often stated that Y-STRs are more informative for detecting recent rather than ancient events in the genetic history of populations, whereas ancient events can be identified more reliably using Y-chromosomal single nucleotide polymorphisms, or SNPs, which have mutation rates 100 000 times lower than Y-STRs (Thomson *et al.*, 2000). One of the first studies describing a geographically restricted distribution of a Y-SNP marker, and its use for investigating human population history, appeared in 1997 (Zerjal *et al.*, 1997). Today, a large number of Y-SNP markers are known and many of them show a continent-specific distribution. A comprehensive summary of the distribution and applications of Y-chromosome SNP markers can be found elsewhere (Jobling *et al.*, 2004). Here, I want to illustrate the suitability of Y-SNP markers for detecting geographical origins using three continent-specific examples:

1. African origins: the Y-SNP marker SRY_{4064} (or one of its phylogenetic equivalents, M96 or P29) defining haplogroup E (Y Chromosome Consortium, 2002) appears at high frequency almost everywhere in Africa but is absent from all regions outside Africa, except those in close geographical proximity to Africa (Plate 9.2a). This is because the mutation probably arose in Africa some time after humans migrated out of Africa, about 150 000 years ago, but before the major human migrations within Africa.

2. European origins: the Y-SNP marker M173 defining the haplogroup R1 (Y Chromosome Consortium, 2002) has a high frequency in Europe (especially Western Europe) and a low to non-existent frequency outside of Europe, except those areas with known records of European immigration, e.g. due to the European colonizations starting about 500 years ago carrying the marker to regions such as Polynesia, together with more recent European admixture, e.g. in New Zealand (Plate 9.2b). This mutation most likely has an ancient origin in Eurasia but its current frequency distribution is believed to be the result of a postglacial expansion starting 20 000–13 000 years ago from a refugee population somewhere on the Iberian peninsula (Semino *et al.*, 2000), explaining the gradual frequency decline from Western to Eastern Europe.

3. East Asian origins: the Y-SNP marker M175 defining the haplogroup O (Y Chromosome Consortium, 2002), has a high frequency in East Asia where it most likely originated, but does not exist elsewhere, except in regions with known East Asian influences, e.g. due to the expansion of Austronesian speakers starting about 6000 years ago in east Asia and carrying the marker to regions such as Polynesia (Kayser *et al.*, 2000a) (Plate 9.2c).

Some cases are known where a high correlation between Y-STRs and Y-SNPs has been observed, such as the statistically significant Y-chromosomal differentiation between Polish and German populations, which is assumed to be a genetic consequence of politically forced population movements during and especially after World War II (Kayser *et al.*, 2005). Furthermore, some European metapopulations, as identified by their Y-STR haplotypes (Roewer *et al.*, 2005), correlate well with specific Y-SNP haplogroups, e.g. the most characteristic Eastern European Y-STR haplotype (17,13,30,25,10,11,13,10–14), together with its close relatives, is associated with Y-SNP haplogroup R1(xR1a1), whereas the most characteristic Western European Y-STR haplotype (14,13,29,24,11,13,13,11–14), together with its close relatives, is associated with Y-SNP haplogroup R1a1

(Kayser *et al.*, 2005). Although in cases of close correlation, Y-STRs and Y-SNP markers alone will reveal the same geographical information, a combination of both marker types can be more informative due to the additional information provided by closely related Y-STR haplotypes (e.g. one-step neighbours) existing on a particular Y-SNP background.

To use any type of genetic marker to identify the geographical origin of an individual, large reference databases are required to establish the marker's geographic distribution. Such a database exists for Y-STR haplotypes, with the YHRD. Unfortunately, however, a similar resource for Y-SNP data is not yet available. Ideally, for Y-chromosome-based geographical origin identification, Y-SNPs and Y-STRs should be combined in a single reference database in order to maximize male-specific ancestry information. Efforts to include Y-SNP data in the YHRD database are currently underway.

There is one severe problem with using Y-chromosomal markers for genetic ancestry identification, namely in cases of individuals with mixed genetic heritage (genetic admixture). For example, a Y chromosome DNA analysis of the son of a European man and an African woman will reveal a European geographical origin in the son in spite of his, most likely, African appearance. A similar analysis in all his male relatives will also reveal European (Y-chromosomal) genetic ancestry, even if all of them produce offspring with African women. In such cases, testing Y-chromosomal markers will be completely misleading if conclusions about a person's appearance are to be drawn from the test results. Continental and population-wide genetic admixture is known from North and South America (Alves-Silva *et al.*, 2000; Kayser *et al.*, 2003) but in principle can be expected in individuals everywhere, with an increased probability in regions with a known history of influences from people of different continental origin, e.g. due to the European expeditions to the Americas and the Pacific, or as a result of the African slave trade to the Americas. Therefore, to reveal a person's geographical origin with a high degree of accuracy, ancestry-informative markers from the Y chromosome need to be combined with those from both mitochondrial DNA (Sigurdsson *et al.*, 2006) and autosomal DNA (Lao *et al.*, 2006).

9.6 The future of Y-chromosomal markers in forensics

Due to the unique properties discussed above and the expected increases in the data content of Y-chromosome reference databases it can be expected that Y-chromosomal markers will be increasingly used in forensic casework in the future, particularly in cases where autosomal STRs do not provide useful information. If an autosomal STR profile can be obtained, e.g. in difficult cases through LCN analysis in combination with laser dissection microscopy, the existing national DNA databases together with essentially individual identification provided by autosomal STRs will always make autosomal STR profiles more informative than those from Y-STRs (given their lack of identifying individuals). However, if no interpretable autosomal STR profile can be obtained from a crime scene sample, Y-STRs are the markers of choice for male lineage identification, including low male and multi-male components in mixed stains.

The forensic use of Y-chromosomal markers for geographical origin identification is also expected to increase but will depend not only on the construction of enlarged worldwide reference databases, but also on non-scientific issues such as the adaptation of national DNA laws or the practical interpretation of existing laws. The forensic application of DNA markers to geographical origin identification is not in agreement with the legislation in those countries where the use of DNA for law enforcement purposes is restricted towards markers that allow DNA identification based on non-coding number codes, as usually obtained from autosomal STRs, and do not allow the use of markers that can reveal other kinds of information. Such countries will have to adjust their legislation if they wish to take advantage of the new scientific possibilities offered by new Y-chromosomal (and other genetic) markers. The Netherlands is one (if not the only) country that has modified its DNA legislation and, since 2003, under specific conditions allows the use of DNA markers for genetic ancestry identification as well as for the identification of externally visible characteristics. However, assumptions about the externally visible characteristics of an individual based on geographical origin identification will always be indirect. Furthermore, the accuracy of the assumptions is highly dependent on the level of correlation between the geographical region and the visible trait and therefore such tests are currently limited to a small number of geographical regions where a high correlation exists. In the future it will be important to understand the genetic basis of externally visible human characteristics, which is challenging due to the complex nature of many genes as well as environmental factors being most likely involved. Such research might provide DNA markers to be used by forensic laboratories as direct predictors of human appearance and thereby help to trace unknown suspects (so far, scientifically possible only for red hair; Grimes *et al.*, 2001).

Finally, the Y-chromosomal markers available today only allow male lineage but not male individual identification. Not being able to differentiate between members of the same male lineage is clearly a major limitation when a Y-chromosomal profile match is obtained (although excluding male individuals is highly valuable too and can be done with a high degree of certainty based on existing Y markers). Future research will determine whether it will be possible to find Y chromosome markers that allow differentiation between close male relatives and thus allow the identification of male individuals and not 'only' groups of male relatives as possible today.

9.7   Acknowledgements

I would like to thank all the colleagues I have had and still have the privilege to collaborate with in the field of Y chromosome research and its applications to forensics and other topics, in particular Peter de Knijff (Leiden, The

Netherlands), Chris Tyler-Smith (Hinxton, UK), Lutz Roewer (Berlin, Germany) and Antti Sajantila (Helsinki, Finland). Chris Tyler-Smith is also acknowledged for sharing the Y-SNP data used to prepare Plate 9.2. I am grateful to Mark Nellist (Rotterdam, The Netherlands) for useful comments on the manuscript.

9.8 References

Akane, A., Shiono, H., Matsubara, K., Nakahori, Y., Seki, S., Nagafuchi, S., Yamada, M. and Nakagome, Y. (1991) Sex identification of forensic specimens by polymerase chain reaction (PCR): two alternative methods. *Forensic Sci. Int.* **49**: 81–88.

Alves-Silva, J., da Silva Santos, M., Guimaraes, P.E., Ferreira, A.C., Bandelt, H.J., Pena, S.D. and Prado, V.F. (2000) The ancestry of Brazilian mtDNA lineages. *Am. J. Hum. Genet.* **67**: 444–461.

Ballard, D.J., Phillips, C., Wright, G., Thacker, C.R., Robson, C., Revoir, A.P. and Court, D.S. (2005) A study of mutation rates and the characterization of intermediate, null and duplicated alleles for 13 Y chromosome STRs. *Forensic. Sci. Int.* **155**: 65–70.

Beleza, S., Alves, C., Gonzalez-Neira, A., Lareu, M., Amorim, A., Carracedo, A. and Gusmao, L. (2003) Extending STR markers in Y chromosome haplotypes. *Int. J. Legal. Med.* **117**: 27–33.

Berta, P., Hawkins, J.R., Sinclair, A.H., Taylor, A., Griffiths, B.L., Goodfellow, P.N. and Fellous, M. (1990) Genetic evidence equating SRY and the testis-determining factor. *Nature* **348**: 448–450.

Betz, A., Bassler, G., Dietl, G., Steil, X., Weyermann, G. and Pflug, W. (2001) DYS STR analysis with epithelial cells in a rape case. *Forensic. Sci. Int.* **118**: 126–130.

Bosch, E. and Jobling, M.A. (2003) Duplications of the AZFa region of the human Y chromosome are mediated by homologous recombination between HERVs and are compatible with male fertility. *Hum. Mol. Genet.* **12**: 341–347.

Budowle, B., Adamowicz, M. Aranda, X.G., Barna, C., Chakraborty, R., Cheswick, D., Dafoe, B., *et al.* (2005) Twelve short tandem repeat loci Y chromosome haplotypes: genetic analysis on populations residing in North America. *Forensic. Sci. Int.* **150**: 1–15.

Budowle, B., Sinha, S., Lee, H.S. and Chakraborty, R. (2003) Utility of Y-chromosome short tandem repeat haplotypes in forensic applications. *Forensic Sc. Rev.* **15**: 154–164.

Butler, J.M., Decker, A.E., Kline, M.C. and Vallone, P.M. (2005) Chromosomal duplications along the Y-chromosome and their potential impact on Y-STR interpretation. *J. Forensic Sci.* **50**: 853–859.

Corach, D., Filgueira Risso, L., Marino, M., Penacino, G. and Sala, A. (2001) Routine Y-STR typing in forensic casework. *Forensic Sci. Int.* **118**: 131–135.

Chakraborty, R. (1985) Paternity testing with genetic markers: are Y-linked genes more efficient than autosomal ones? *Am. J. Med. Genet.* **21**: 297–305.

Daiger, S.P. and Chakraborty, R. (1985) Mapping of the human Y chromosome. In: *The Cytogenetics of the Mammalian Y Chromosomes* (A. Sandberg, ed.), Alan R. Liss, New York, pp. 1–31.

de Knijff, P. (2003) Son, give up your gun: presenting Y-STR results in court. *Profiles DNA* **6**: 3–5.

Dettlaff-Kakol, A. and Pawlowski, R. (2002) First Polish DNA 'manhunt' – an application of Y-chromosome STRs. *Int. J. Legal. Med.* **116**: 289–291.

Dupuy, B.M., Stenersen, M., Egeland, T. and Olaisen, B. (2004) Y-chromosomal microsatellite mutation rates: differences in mutation rate between and within loci. *Hum. Mutat.* **23**: 117–124.

Epplen, J.T., Cellini, A., Romero, S. and Ohno, S. (1983) An attempt to approach the molecular mechanism of primary sex determination: W- and Y-chromosomal conserved simple repetitive sequences and their differential expression in mRNA. *J. Exp. Zool.* **228**: 305–312.

Foster, E.A., Jobling, M.A., Taylor, P.G., Donnelly, P., de Knijff, P., Mieremet, R., Zerjal, T. and Tyler-Smith, C. (1998) Jefferson fathered slave's last child. *Nature* **396**: 27–28.

Gill, P., Brenner, C., Brinkmann, B., Budowle, B., Carracedo, A., Jobling, M.A., *et al.* (2001) DNA Commission of the International Society of Forensic Genetics: recommendations on forensic analysis using Y-chromosome STRs. *Forensic Sci. Int.* **124**: 5–10.

Grimes, E.A., Noake, P.J., Dixon, L. and Urquhart, A. (2001) Sequence polymorphism in the human melanocortin 1 receptor gene as an indicator of the red hair phenotype. *Forensic Sci. Int.* **122**: 124–129.

Gusmao, L., Butler, J.M., Carracedo, A., Gill, P., Kayser, M., Mayr, W.R., Morling, N. Prinz, M., Roewer, L., Tyler-Smith, C. and Schneider, P.M. (2006) DNA Commission of the International Society of Forensic Genetics (ISFG): an update of the recommendations on the use of Y-STRs in forensic analysis. *Forensic Sci. Int.* **157**: 187–197.

Gusmao, L., Sanchez-Diz, P., Calafell, F., Martin, P., Alonso, C.A., Alvarez-Fernandez, F., *et al.* (2005) Mutation rates at Y chromosome specific microsatellites. *Hum. Mutat.* **26**: 520–528.

Hall, A. and Ballantyne, J. (2003) Novel Y-STR typing strategies reveal the genetic profile of the semen donor in extended interval post-coital cervicovaginal samples. *Forensic Sci. Int.* **136**: 58–72.

Hedman, M., Pimenoff, V., Lukka, M., Sistonen, P. and Sajantila, A. (2004) Analysis of 16 Y STR loci in the Finnish population reveals a local reduction in the diversity of male lineages. *Forensic Sci. Int.* **142**: 37–43.

Heyer, E., Puymirat, J., Dieltjes, P., Bakker, E. and de Knijff, P. (1997) Estimating Y chromosome specific microsatellite mutation frequencies using deep rooting pedigrees. *Hum. Mol. Genet.* **6**: 799–803.

Jacobs, P.A. and Strong, J.A. (1959) A case of human intersexuality having a possible XXY sex-determining mechanism. *Nature* **183**: 302–303.

Jobling, M.A., Heyer, E., Dieltjes, P. and de Knijff, P. (1999) Y-chromosome-specific microsatellite mutation rates re-examined using a minisatellite, MSY1. *Hum. Mol. Genet.* **8**: 2117–2120.

Jobling, M.A., Hurles, M.E. and Tyler-Smith, C. (2004) *Human Evolutionary Genetics: Origins, Peoples, and Disease.* Garland Science Taylor & Francis Group, New York.

Jones, A.W. (2007) The distribution of forensic journals, reflections on authorship practices, peer-review and role of the impact factor. *Forensic Sci. Int.* **165**: 115–128.

Kayser, M., Brauer, S., Schadlich, H., Prinz, M., Batzer, M.A., Zimmerman, P.A., Boatin, B.A. and Stoneking, M. (2003) Y chromosome STR haplotypes and the genetic structure of U.S. populations of African, European, and Hispanic ancestry. *Genome Res.* **13**: 624–634.

Kayser, M., Brauer, S., Weiss, G., Underhill, P.A., Roewer, L., Schiefenhovel, W. and Stoneking, M. (2000a) Melanesian origin of Polynesian Y chromosomes. *Curr. Biol.* **10**: 1237–1246.

Kayser, M., Brauer, S., Willuweit, S., Schadlich, H., Batzer, M.A., Zawacki, J., Prinz, M., Roewer, L. and Stoneking, M. (2002) Online Y-chromosomal short tandem repeat haplotype reference database (YHRD) for U.S. populations. *J. Forensic Sci.* **47**: 513–519.

Kayser, M., Caglia, A., Corach, D., Fretwell, N., Gehrig, C., Graziosi, G., *et al.* (1997) Evaluation of Y-chromosomal STRs: a multicenter study. *Int. J. Legal Med.* **110**: 125–133, 141–149.

Kayser, M., Kittler, R., Erler, A., Hedman, M., Lee, A.C., Mohyuddin, A., *et al.* (2004) A comprehensive survey of human Y-chromosomal microsatellites. *Am. J. Hum. Genet.* **74**: 1183–1197.

Kayser, M., Krawczak, M., Excoffier, L., Dieltjes, P., Corach, D., Pascali, V., *et al.* (2001) An extensive analysis of Y-chromosomal microsatellite haplotypes in globally dispersed human populations. *Am. J. Hum. Genet.* **68**: 990–1018.

Kayser, M., Krueger, C., Nagy, M., Geserick, G., de Knijff, P. and Roewer, L. (1998) Y-chromosomal DNA-analysis in paternity testing: experiences and recommendations. In: *Progress in Forensic Genetics*, vol. 7 (B. Olaisen, B. Brinkmann and P.J. Lincoln, eds), Elsevier Science, Amsterdam, pp. 494–496.

Kayser, M., Lao, O., Anslinger, K., Augustin, C., Bargel, G., Edelmann, J., *et al.* (2005) Significant genetic differentiation between Poland and Germany follows present-day political borders, as revealed by Y-chromosome analysis. *Hum. Genet.* **117**: 428–443.

Kayser, M., Roewer, L., Hedman, M., Henke, L., Henke, J., Brauer, S., *et al.* (2000b) Characteristics and frequency of germline mutations at microsatellite loci from the human Y chromosome, as revealed by direct observation in father/son pairs. *Am. J. Hum. Genet.* **66**: 1580–1588.

Kayser, M. and Sajantila, A. (2001) Mutations at Y-STR loci: implications for paternity testing and forensic analysis. *Forensic Sci. Int.* **118**: 116–121.

Kiel-Metzger, K., Warren, G., Wilson, G.N. and Erickson, R.P. (1985) Evidence that the human Y chromosome does not contain clustered DNA sequences (BKM) associated with heterogametic sex determination in other vertebrates. *N. Engl. J. Med.* **313**: 242–245.

Koopman, P., Gubbay, J., Vivian, N., Goodfellow, P. and Lovell-Badge, R. (1991) Male development of chromosomally female mice transgenic for Sry. *Nature* **351**: 117–121.

Krawczak, M. (2001) Forensic evaluation of Y-STR haplotype matches: a comment. *Forensic Sci. Int.* **118**: 114–115.

Lao, O., van Duijn, K., Kersbergen, P., de Knijff, P. and Kayser, M. (2006) Proportioning whole-genome single-nucleotide-polymorphism diversity for the identification of geographic population structure and genetic ancestry. *Am. J. Hum. Genet.* **78**: 680–690.

Lessig, R., Willuweit, S., Krawczak, M., Wu, F.C., Pu, C.E., Kim, W., *et al.* (2003) Asian online Y-STR Haplotype Reference Database. *Legal Med. (Tokyo)* **5** (Suppl. 1): 160–163.

Mohyuddin, A., Ayub, Q., Qamar, R., Zerjal, T., Helgason, A., Mehdi, S.Q. and Tyler-Smith, C. (2001) Y-chromosomal STR haplotypes in Pakistani populations. *Forensic Sci. Int.* **118**: 141–146.

Nuzzo, F., Caviezel, F. and De Carli, L. (1966) Y chromosome and exclusion of paternity. *Lancet* **2**: 260–262.

Painter, T. (1923) Studies in mammalian spermatogenesis. II. The spermatogenesis of man. *J. Exp. Zool.* **37**: 291–335.

Park, M.J., Lee, H.Y., Yoo, J.E., Chung, U., Lee, S.Y. and Shin, K.J. (2005) Forensic evaluation and haplotypes of 19 Y-chromosomal STR loci in Koreans. *Forensic Sci. Int.* **152**: 133–147.

Pearson, P.L. and Bobrow, M. (1970) Definitive evidence for the short arm of the Y chromosome associating with the X chromosome during meiosis in the human male. *Nature* **226**: 959–961.

Pereira, L., Prata, M.J. and Amorim, A. (2002) Mismatch distribution analysis of Y-STR haplotypes as a tool for the evaluation of identity-by-state proportions and significance of matches – the European picture. *Forensic Sci. Int.* **130**: 147–155.

Prinz, M., Boll, K., Baum, H. and Shaler, B. (1997) Multiplexing of Y chromosome specific STRs and performance for mixed samples. *Forensic Sci. Int.* **85**: 209–218.

Radam, G. and Strauch, H. (1973) Lumineszenzmikroskopischer Nachweis des Y Chromosoms in Knochenmarkzellen – eine neue Methode zur Geschlechtsbestimmung an Leichenmaterial. *Kriminal. Forens. Wissensch.* **6**: 149–151.

Redd, A.J., Agellon, A.B., Kearney, V.A., Contreras, V.A., Karafet, T., Park, H., de Knijff, P., Butler, J.M. and Hammer, M.F. (2002) Forensic value of 14 novel STRs on the human Y chromosome. *Forensic Sci. Int.* **130**: 97–111.

Roewer, L., Arnemann, J., Spurr, N.K., Grzeschik, K.H. and Epplen, J.T. (1992) Simple repeat sequences on the human Y chromosome are equally polymorphic as their autosomal counterparts. *Hum. Genet.* **89**: 389–394.

Roewer, L., Croucher, P.J., Willuweit, S., Lu, T.T., Kayser, M., Lessig, R., de Knijff, P., Jobling, M.A., Tyler-Smith, C. and Krawczak, M. (2005) Signature of recent historical events in the European Y-chromosomal STR haplotype distribution. *Hum. Genet.* **116**: 279–291.

Roewer, L. and Epplen, J.T. (1992) Rapid and sensitive typing of forensic stains by PCR amplification of polymorphic simple repeat sequences in case work. *Forensic Sci. Int.* **53**: 163–171.

Roewer, L., Kayser, M., de Knijff, P., Anslinger, K., Betz, A., Caglia, A., *et al.* (2000) A new method for the evaluation of matches in non-recombining genomes: application to Y-chromosomal short tandem repeat (STR) haplotypes in European males. *Forensic Sci. Int.* **114**: 31–43.

Roewer, L., Kayser, M., Dieltjes, P., Nagy, M., Bakker, E., Krawczak, M. and de Knijff, P. (1996) Analysis of molecular variance (AMOVA) of Y-chromosome-specific microsatellites in two closely related human populations. *Hum. Mol. Genet.* **5**: 1029–1033.

Roewer, L., Krawczak, M., Willuweit, S., Nagy, M., Alves, C., Amorim, A., *et al.* (2001) Online reference database of European Y-chromosomal short tandem repeat (STR) haplotypes. *Forensic Sci. Int.* **118**: 106–113.

Rolf, B., Keil, W., Brinkmann, B., Roewer, L. and Fimmers, R. (2001) Paternity testing using Y-STR haplotypes: assigning a probability for paternity in cases of mutations. *Int. J. Legal Med.* **115**: 12–15.

Santos, F.R., Pandya, A. and Tyler-Smith, C. (1998) Reliability of DNA-based sex tests. *Nat. Genet.* **18**: 103.

Semino, O., Passarino, G., Oefner, P.J., Lin, A.A., Arbuzova, S., Beckman, L.E., *et al.* (2000) The genetic legacy of Paleolithic Homo sapiens sapiens in extant Europeans: a Y chromosome perspective. *Science* **290**: 1155–1159.

Sigurdsson, S., Hedman, M., Sistonen, P., Sajantila, A. and Syvanen, A.C. (2006) A microarray system for genotyping 150 single nucleotide polymorphisms in the coding region of human mitochondrial DNA. *Genomics* **87**: 534–542.

Sinclair, A.H., Berta, P., Palmer, M.S., Hawkins, J.R., Griffiths, B.L., Smith, M.J., Foster, J.W., Frischauf, A.M., Lovell-Badge, R. and Goodfellow, P.N. (1990) A gene from the human sex-determining region encodes a protein with homology to a conserved DNA-binding motif. *Nature* **346**: 240–244.

Stalvey, J.R. and Erickson, R.P. (1987) An improved method for detecting Y chromosomal DNA. *Hum. Genet.* **76**: 240–243.

Steinlechner, M., Berger, B., Niederstatter, H. and Parson, W. (2002) Rare failures in the amelogenin sex test. *Int. J. Legal Med.* **116**: 117–120.

Thangaraj, K., Reddy, A.G. and Singh, L. (2002) Is the amelogenin gene reliable for gender identification in forensic casework and prenatal diagnosis? *Int. J. Legal Med.* **116**: 121–123.

Thomson, R., Pritchard, J.K., Shen, P., Oefner, P.J. and Feldman, M.W. (2000) Recent common ancestry of human Y chromosomes: evidence from DNA sequence data. *Proc. Natl. Acad. Sci. USA* **97**: 7360–7365.

Turrina, S., Atzei, R. and De Leo, D. (2006) Y-chromosomal STR haplotypes in a Northeast Italian population sample using 17plex loci PCR assay. *Int. J. Legal Med.* **120**: 56–59.

Witt, M. and Erickson, R.P. (1989) A rapid method for detection of Y-chromosomal DNA from dried blood specimens by the polymerase chain reaction. *Hum. Genet.* **82**: 271–274.

Y Chromosome Consortium (2002) A nomenclature system for the tree of human Y-chromosomal binary haplogroups. *Genome Res.* **12**: 339–348.

Zerjal, T., Dashnyam, B., Pandya, A., Kayser, M., Roewer, L., Santos, F.R., *et al.* (1997) Genetic relationships of Asians and Northern Europeans, revealed by Y-chromosomal DNA analysis. *Am. J. Hum. Genet.* **60**: 1174–1183.

10

Laser microdissection in forensic analysis

Luigi Saravo, Davide Di Martino, Nicola Staiti, Carlo Romano, Enrico Di Luise, Dario Piscitello, Salvatore Spitaleri, Ernesto Ginestra, Ignazio Ciuna, Fabio Quadrana, Beniamino Leo, Giuseppe Giuffrè and Giovanni Tuccari

10.1   Introduction

Forensic science is aimed at detecting and analysing evidential material at crime scenes. Collection and careful laboratory analysis of any trace material is an extremely important activity in order to obtain as many pieces of information as possible. Considerable skill is involved in conducting searches at crime scenes. Frequently, particularly where sexual assault has occurred, crime scene investigators encounter mixed biological traces. The ability to isolate different cell populations (i.e. vaginal/sperm cells, epithelial/white cells, white/sperm cells, etc.) is therefore desirable to ensure the successful outcome of an investigation (Murray and Curran, 2005). Laser microdissection techniques have recently been introduced into forensic analysis (Giuffrè *et al.*, 2005) and have proved to be very powerful in the isolation of specific cells from complex mixtures of biological material (Giuffrè *et al.*, 2004). It has also increased the types of samples that can be collected from crime scenes and usefully analysed (Plate 10.1).

Laser microdissection techniques fall within the field of microgenomics, which has been referred to as 'a quantitative molecular analysis of nucleic acids or proteins obtained from a single cell or a tiny amount of cells, which were isolated, collected and examined according to precise micromanipulating techniques'.

Molecular Forensics.   Edited by Ralph Rapley and David Whitehouse
Copyright 2007 by John Wiley & Sons, Ltd.

Micro-isolation and micromanipulation techniques are essential when undertaking laser microdissection techniques. Laser capture and laser cutting are the main techniques applied to perform cell micro-isolation.

Emmert-Buck and colleagues (1996) developed the first laser capture microdissection device, which was patented by Arcturus in the following year (Plate 10.2). It was based upon the heating and fusion of a thermo-sensitive plastic polymer that is modified by low-energy laser pulses (with near-infrared wavelength) (Plate 10.3). This is also a useful device when dealing with RNA because of its sensitivity to high temperatures. However, the drawback is that the method lacks sufficient precision to isolate target cells from the matrix (Plate 10.4).

Laser cutting is preferable because the precision of the laser allows the isolation of single cells or a group of cells from more complex biological matrices such as tissue sections or hair shafts. Table 10.1 compares the main features of laser cutting and laser capture microdissection techniques.

Modern laser cutting devices exploit a solid-state UV laser beam with a wavelength between 337 nm and 370 nm. Such a beam is able to behave like an electromagnetic knife and can destroy whatever it meets, including the biological substrate where cells are usually found. The cutting is done with great precision down to a single micrometre.

The cell sample is focused with the aid of a microscope through which the operator can select the target and then cut using the laser beam. The three main laser cutting systems may be defined by the collection method used:

- Collection by gravity.

- Collection by an adhesive polymer applied on a tube cap.

- Collection through catapulting or energetic pulse.

Laser microdissecting instruments are coupled to different microscope systems depending on the collection method to be used. For example, a conventional upright microscope may be used for gravity-based collection, while an inverted microscope is used for the other two systems of collection.

Table 10.1 Comparison of laser capture microdissection and laser cutting techniques

Laser capture microdissection	Laser cutting
Infrared beam	Ultravidet beam
Suitable for isolation of a small number of cells	Suitable for wide sections
Not harmful to nearby tissues/cells	Destroys where it cuts
Not appropriate for hard target tissues	Appropriate for hard target tissues

In the case of gravity-based collection the operator can observe the target cells through the microscope at an appropriate magnification that enables the selection and cutting of the target with the aid of specific imaging software. The last laser pulse will allow the target sample to be collected under gravity into the tube cap placed immediately under the microscope. The successful removal of the sample can be confirmed by re-focusing the microscope onto the tube cap (Plate 10.5). A disadvantage of this system is that it does not allow for the possibility that the wrong sample can be cut off, which would be extremely important in forensic medicine where non-repeatable analyses are carried out.

The micro-isolation device based on the adhesive polymer allows visualization of the target sample through an inverted microscope and then, after cutting, collection by direct contact between the specimen and the tube cap where the polymer is located; in this way the target sample is torn away from the specimen. The adhesiveness of the polymer ensures that the sample is removed and its presence is confirmed by microscopy. A disadvantage of this system is that the contact between the slide and the cap could cause transfer of contaminating material. With catapulting microdissection, an inverted microscope is coupled to a high-resolution CCD camera showing the collected sample on the collection tube cap where a laser pulse has pushed it (Plate 10.6). There is no contact between specimen and collection tube and hence no need for any adhesive polymer to retain the cut sample, which can be easily visualized by the microscope. Moreover, the collection tube is positioned within a few microns of the specimen from which the target sample will be cut, which makes sample collection a very precise and contamination-free activity.

The choice of method will depend on the specific biological problem under investigation and individual preferences. The choice of microdissecting device/ collection method is the starting point and there are several additional steps that must be taken into consideration in order to achieve good results in forensic medicine. These are summarized as follows:

- Requirement for a dedicated 'isolated' laboratory exclusively for the use of microdissecting activity.

- Requirement for personal protective equipment (PPE).

- Specific collection tubes that are sterilized and, if possible, made of low-binding plastic.

- Ultraviolet-sterilized environment.

- Isolated air supply in microdissection room.

- Requirement for humidity and temperature monitoring.

- Rigorous quality control of instrumentation.

- Dedicated histological reagents.

- DNA typing of all operators involved in the analysis chain.

- Dissection and processing of control material not concurrent with target sample.

- Performing analysis on duplicate samples if possible.

10.2   Histological, biochemical analysis

The aim of laser micro-isolating applications is to isolate specific nucleated cells and extract DNA for STR typing in order to determine a DNA profile (Plate 10.7). The first step is to make a cell smear on a microscope glass in order to visualize the specimen under the microscope lens. The method for specimen preparation depends on the nature of the biological material to be analysed (Di Martino *et al.*, 2004a, 2004b). In any case, the material of interest must be collected on a supporting synthetic polymeric membrane – polyethylene tereph-thalate (PET) or polyethylene naphthalate (PEN) – that may be mounted onto a glass slide or a metallic frame. In this way, during the laser cutting of an area containing a distinct cell type, a fragment of supporting membrane is collected together with the sample of interest into the microtube. The presence of the membrane does not interfere with the analytical procedures. Prior to specimen mounting, in order to avoid nucleic acid contaminations of the membrane, all procedures take place in a dedicated DNA-free area, and the membrane is steri-lized by autoclaving and/or UV treatment. The latter is particularly useful because it eliminates electrostatic charges from the membrane, thus avoiding adhesive effects on the laser-cut fragment that could interfere with harvesting the cells.

In forensic analysis, a wide range of material, such as blood, organic fluids, hair or a mixture, may be collected on the membrane and subsequently subjected to morphological identification by laser-based methods. Specific identification of cell type, such as spermatozoa, epithelial cells from mucosae or skin, leuco-cytes or cells from hair follicles, frequently requires the use of several micro-scopic procedures.

The cell smear can be stained or left unstained. In the former case the opera-tor must employ a stain that will not damage nuclear DNA, whereas in the latter case phase-contrast microscopy is used. It is often preferable to work on stained samples if the forensic laboratory is supported by a histological service that can perform the most appropriate cytological/histological staining. Good slide preparation and cellular morphology is essential to correctly distinguish different cell types. It will also ensure that the quality of DNA is sufficient to produce an unambiguous result. Many staining procedures have been investigated, leading to the identification of Papanicolau, Haematoxylin/Eosin and Giemsa as the basis of the best staining protocols for forensic applications. For the great majority of specimens, the optimal balance between morphology and DNA integrity is usually obtained by air fixation and Giemsa staining (Di Martino *et al.*, 2004a). The above-mentioned stains do not intercalate DNA nor are they able to fragment the DNA backbone. Importantly they do not cause polymerase chain reaction (PCR) inhibition either. Other chemical staining agents, such as Nuclear Fast Red-Picroindigocarmine, may influence DNA preservation in spermatozoa and result in poor single random repeat (STR) DNA profiles. No matter which stain is used, in order to avoid contamination all the staining steps must be performed by dispensing the reagents on the surface of the specimen, which is then placed horizontally into an incubation chamber. Before starting the microdissection procedure, it is very important to verify that the specimen is dry, especially when membrane-coated slides are utilized, because any moisture present between the glass surface and the membrane may interfere with the detachment of the cut portion. An important step shared by all microdissection procedures is inspection of the cap of the microtube to confirm that the cut fragment has been collected. To facilitate the identification of membrane fragments, it is useful to leave a margin around the cell, drawing an irregular distinguishable outline.

Extraction of DNA from cell samples collected by laser microdissection is another delicate step. Different extraction protocols will be required according to the kind of specimen the operator is dealing with. Cell types differ in their biochemical and ultrastuctural properties: for example, amounts of fatty acids, phospholipids and proteins, the presence or absence of a cell wall, etc. Certain types of cell are more resistant to lysis, which is the primary step in DNA extraction. Lysis/DNA extraction takes place in a pH-controlled aqueous buffer with a hydrophobic environment useful for membrane disruption and in the presence of a protease.

Because forensic specimens are often typified by very low concentrations of potentially degraded DNA, consideration must be given to different extraction protocols at the outset of each case so that the best method can be employed (Di Martino *et al.*, 2004b; Giuffrè *et al.*, 2004).

A wide range of extraction kits/products are commercially available. Not all are suited to laser microdissected samples because of the small physical dimensions of the samples, the small amounts of DNA and the presence of PCR inhibitors, etc. Three types of reagents have emerged as being most suited to DNA extraction from laser microdissected samples (Chelex®, ion exchange chromatography and magnetic resin) and modified protocols have been developed in all cases.

1. Chelex®: a resin capable of chelating bivalent ions present in the cell lysate; such ions are potential PCR inhibitors and nuclease cofactors. The advantage of this method is that the whole lysate may be treated without further cleavage, although this is not a purification system. The concentration of the resin solution depends on the kind of cell sample under investigation, while the amount of resin required is usually limited in the case of laser microdissected samples. The extraction volume is related to the number of microdissected cells and can be decreased to a few microlitres in the case of DNA extraction from a single cell (Staiti *et al.*, 2005).

2. Ion exchange chromatography: this is a good purification system that can be used for some difficult samples. It is based on alkaline lysis and the separation of cellular debris from DNA using an ion-exchange silica column. Its application to laser microdissected samples is dependent on the number of cells available (down to 8–10 haploid cells), but it is not so useful when DNA extraction is performed on five cells or less. A potential drawback is that DNA damage could occur as a result of the alkaline environment, which is very important when working in low copy number conditions. DNA extracted from 1–10 microdissected cells needs to be dissolved in very small volumes of buffer solution because it will greatly influence the quality of the subsequent PCR.

3. Magnetic resin: a purification system based upon the separation of cell debris from DNA by magnetic beads suspended in the same solution as the cell lysate. The chemistry of the system is most suited to laser microdissected spermatozoa and hair bulbs. As a consequence of the buffer composition, complete recovery when working with a very restricted number of cells (one or two) is seldom possible. Moreover, it is a very challenging method that requires great operating skill, otherwise DNA may become degraded. This has been noticed especially when dealing with telogen laser microdissected hair bulbs, where old and totally keratinized cells are present and their DNA is already partially fragmented.

Laser microdissection allows operators to isolate specific cells from minute samples to determine DNA profiles, gender and species of origin (Di Martino *et al.*, 2005). As laser microdissection generates DNA templates at the lower limits of concentration, it is essential to consider the stoichiometry of the PCR.

There are two PCR strategies in general use. First, DNA can be separated into several aliquots and individually amplified. This is the traditional method of nucleic acid amplification, which can be applied to laser microdissected samples. This approach allows the detection of allelic drop-out since multiple aliquots are available for analysis. Second, the operator performs a single reaction of amplification, using all the DNA solution extracted from the forensic specimen. In principle any allelic drop-out observed is likely to be a biologically related phenomenon rather than an artefact of the PCR (Plate 10.8).

An intermediate situation between a complete STR-typed allelic profile and one affected by allelic drop-out can occur as a consequence of the reduced amount of DNA template in the PCR solution. When examining genotype peak heights it is important to discriminate, within an apparent heterozygous genotype, whether unbalanced alleles are due to the presence of a stutter product at that locus. Only a precise determination of allele peak areas or, even better, an initial design of a duplicate reaction set for each sample can assist in this situation.

In summary, operators aiming to perform analysis of low copy number DNA by the laser microdissection technique should take into account the following guidelines:

1. Adjust the protocols in order to work on low volumes.

2. Reach the maximum grade of purity of the extracted DNA; additionally evaluate the possibility to increase the injection time during capillary electrophoresis in order to reduce the influence of low-molecular-weight particles.

3. Duplicate the PCR assay on the same sample in order to potentially minimize any stochastic effect.

4. Avoid unnecessarily increasing the number of PCR cycles; the *Taq* polymerase concentration can be increased within the PCR reaction mix.

10.3 References

Di Martino, D., Giuffrè, G., Staiti, N., Simone, A., Le Donne, M. and Saravo, L. (2004a) Single sperm cell isolation by laser microdissection. *Forensic Sci. Int.* **146**: S151–S153.

Di Martino, D., Giuffrè, G., Staiti, N., Simone, A., Sippelli, G., Tuccari, G. and Saravo, G. (2005) LMD as a forensic tool in a sexual assault casework: LCN DNA typing to identify the responsible. *Progr. Forensic Genet.* **11**: 571–573.

Di Martino, D., Giuffrè, G., Staiti, N., Simone, A., Todaro, P. and Saravo, L. (2004b) Laser microdissection and DNA typing of cells from single hair follicles. *Forensic Sci. Int.* **146**: S155–S157.

Elliot, K., Hill, D.S., Lambert, C., Burroughes, T.R. and Gill, P. (2003) Use of laser microdissection greatly improves the recovery of DNA from sperm on microscope slides. *Forensic Sci. Int.* **137**: 28–36.

Emmert-Buck, M.R., Bonner, R.F., Smith, P.D., Chuaqui, R.F., Zhuang, Z., Goldstein, S.R., Weiss, R.A. and Liotta, L.A. (1996) Laser capture microdissection. *Science* **274**: 998–1001.

Giuffrè, G., Saravo, L., Di Martino, D., Staiti, N., Simone, A., Todaro, P. and Tuccari, G. (2004) Analyses of genomic low copy number DNA from cells harvested by laser microdissection. *Pathologica* **96**: 396.

Giuffrè, G., Saravo, L., Di Martino, D., Staiti, N., Simone, A., Sippelli, G. and Tuccari, G. (2005) Molecular analyses of genomic low copy number DNA extracted from laser-microdissected cells. *Virch. Arch.* **447**: 515.

Murray, G.I. and Curran, S. (2005) *Laser Capture Microdissection – Methods and Protocols.* Humana Press, Totowa, NJ.

Staiti, N., Giuffrè, G., Di Martino, D., Simone, A., Sippelli, V., Tuccari, G. and Saravo, L. (2005) Molecular analysis of genomic low copy number DNA extracted from laser-microdissected cells. *Progr. Forensic Genet.* **11**: 568–570.

11

Laboratory information systems for forensic analysis of DNA evidence

Benoît Leclair and Tom Scholl

11.1 Introduction

Since their introduction into forensics some two decades ago, DNA-based geno-typing technologies have revolutionized the science of human identification. Initially, panels of 'variable number tandem repeats' (VNTR) probes were used in Southern blot applications and represented, at the time, a quantum leap over serological testing in their ability to discriminate between individuals. The invention of the polymerase chain reaction (PCR) fostered the introduction of the much shorter hypervariable 'short tandem repeats' (STR), the major forensic DNA typing tool currently used, as well as mitochondrial DNA sequencing and, more recently, Y-STRs and single nucleotide polymorphisms (SNPs). Compared to VNTR analysis, PCR-based assays are better suited for the often-compromised nature of crime scene samples, they consume 1000-fold less sample and they reduce the sample processing time from weeks to less than 24 hours. Despite these significant advances in analytical methods, improvements in throughput have largely been offset by increased specimen collection at crime scenes and through expansion of legislated mandates for criminal offences. Although the existence of backlogs in many jurisdictions may be associated with high local crime rates, the resource-consuming, manual nature of forensic casework as it is performed in most laboratories remains a significant factor contributing

Molecular Forensics.   Edited by Ralph Rapley and David Whitehouse
Copyright 2007 by John Wiley & Sons, Ltd.

to the backlogs. Additional opportunities for improving processing capacity and turn-around time reside in the development of faster evidence screening tools and in the introduction of automation for liquid handling and data analysis.

The elimination of offender and casework backlogs and the timely processing of incoming submissions are necessary steps towards achieving the true potential of DNA typing technologies. No less necessary is the development of informatic tools not only to provide support for automation technologies necessary for improvements in process quality, reliability and throughput, but also to take full advantage of the information content of processed data sets. Although the resolution of many casework situations resides in direct matching of genotypes obtained from crime scene evidence and suspect(s), applications requiring more intricate data analysis have emerged in response to challenges of increasing complexity encountered in the field. Current laboratory information system (LIS) developments on the offender data banking front have largely dealt with supporting parallel processing pathways and automated data review to increase the reliability of uploaded offender genotype data. Other activities on that front aim at extending direct match capabilities to 'familial' search capabilities, as many offenders have been found to have next-of-kin already included in offender data banks. Data analysis applications are emerging on the crime scene front to assist with data interpretation of mixed profiles that often necessitate software-assisted, mathematical deconvolution for resolution. In situations where recovered human remains must be identified, significantly different approaches are required because a reference biological sample from the victim is often not available. In mass fatality incidents (MFIs), entire families often perish, making it necessary to use large-scale kinship analysis to re-construct family pedigrees from within the victim's genotypic data set. The LIS applications have been built to address these numerous complexities on the path to identification in large-scale MFIs and ongoing developments aim at providing computing capabilities for even larger scale and more complex events. Much of the LIS infrastructure used for MFIs can be leveraged into the development of Missing Persons Databasing applications as very similar complexities are encountered.

This chapter presents the developments and applications of laboratory information systems that have promoted the growth of genetic forensic identification over the last decade. First, the major DNA analytical platform used in forensics is discussed to identify areas where automation technologies can provide improvements in the platform's qualitative and quantitative performance. Next, specific implementations of computing support to different forensic applications are reviewed. Finally, conclusions are presented regarding likely future directions for the adoption of technology and development activities by forensic laboratories.

11.2 The specifications of forensic genotyping assays

The choice of genetic markers and assay technologies needs to be carefully considered if these technologies are to be employed in forensic genotype data banking. A decade ago, the forensic science community recognized in autosomal STRs a general genotyping technology with the potential for responding to data banking and casework needs (Kimpton *et al.*, 1993; Urquhart *et al.*, 1994; Lygo *et al.*, 1994). Since then, forensic laboratory facilities have developed considerable experience and invested heavily in an installed base of this technology. The platform has proven robust and cost-effective, performs well in a variety of operational contexts and, importantly, allows for casework mixed-source samples to be analysed. Although many other genetic markers, e.g. mtDNA (Hagelberg *et al.*, 1991; Sullivan *et al.*, 1992), SNPs (Gill, 2001; Gill *et al.*, 2004) and Y-STRs (Hall and Ballantyne, 2003; Schoske *et al.*, 2004), have been developed and are put to use in specific operational contexts, the current autosomal STR system is likely to remain the reference platform for forensic DNA analysis platforms for the foreseeable future (Jobling and Gill, 2004). This will not impede development work on wet chemistry components of this assay as throughput improvements are expected to emerge from the implementation of thermal cycling and electrophoresis capabilities embarked on micro-fabricated devices.

Single tandem repeat genotyping is accomplished through multiplex PCR reactions designed to amplify up to 16 genetic markers or 32 different allelic targets in a single reaction. Its range of alleles and the presence of 1 bp variants confer high discrimination potential to the system and can be resolved with electrophoresis instrumentation designed for DNA sequencing applications. Several multiplexes configured for forensic application are commercially available and produce electropherograms balanced for intra- and inter-locus signal strength over a variety of forensic sample types. Precision and accuracy in allele calling are achieved on electrophoresis platforms with floating bins through the use of commercially available allelic ladders. A variable and generally small percentage of samples will either experience some random processing anomaly or present a rare but normal feature, both of which will be detected on electropherograms. These anomalies (e.g. 'spike', pull-up, saturated and split peaks, elevated baseline, heterozygote ratio imbalance, elevated stutter, profile slope), at times, may interfere with accurate allele calling, or cause the affected sample to fall outside of quality control specifications. Rare features may prove to be variant alleles, tri-allelic loci or unfavourable heterozygous ratio at a given locus as a consequence of a sequence polymorphism under a primer annealing site. The interpretation uncertainty associated with both random processing anomalies and rare genetic features is normally resolved by re-working the sample through the analytical platform: the latter condition will be replicated, the former will be resolved.

From an automation design perspective, the STR technology is stable – a desirable attribute as it limits the scope and costs of software maintenance/ upgrades. The labour-intensive procedure involved in STR typing offers substantial opportunities for quality, reliability and productivity improvements through the introduction of automated solutions.

11.3 Automated pipetting

As with any analytical platform, manual pipetting is best suited for the handling of very few samples at a time, and quickly becomes error-prone as throughput escalates. Automated liquid handling is the ideal solution in large-sample volume applications, such as convicted offender data banking. Proven robotic technologies tested in genome sequencing and clinical environments can be readily implemented in forensic environments, and their application can provide benefits in process precision and reproducibility, increased sample tracking confidence through reduction of error-prone tasks, the reliability of movements generated from computer programs and sample throughput.

Sample tracking confidence is an important aspect of data quality that automated pipetting can improve compared with manual processes. High-density assay plates are required for high-throughput applications involving the low reaction volumes of PCR-based assays, an environment where the high positional accuracy and reliability of automated pipetters are considered a prerequisite to secure sample tracking. The majority of robotic instruments can integrate barcode scanning devices to record the location and identity of all barcoded containers on the instrument work surfaces, assuring correct container addressing by pipetting or transport heads. A powerful feature available through this enhancement is operator-independent sample and reagent tracking. Many of these robotic systems offer logging capabilities that support quality assurance and chain-of-custody objectives by recording pipetting steps associated with specimen processing.

More sophisticated implementations support communication with external systems such as LIS applications (Scholl *et al.*, 1998; Frégeau *et al.*, 2003; Leclair *et al.*, 2004b; Scholl, 2004; Leclair and Scholl, 2005). Logic can be introduced to permit the LIS to control many aspects of sample flow management. Client-based applications can be designed to initiate a transaction containing information about specimens, reagents, operators, instruments, etc. prior to initiating a processing step. After the LIS has confirmed that sequential processing steps are completed in order, reagents are appropriate and have passed quality control tests, operators and equipment are validated, as well as making other important confirmations, a transaction is returned to the client application to signal that processing can proceed. A final transaction is sent back by the client application to the database to record that the process is complete. Since queries to the database can be incorporated at the outset of various processing steps, the LIS can track the progress of every sample, prevent inappropriate processing and alert operators about detected incongruities.

A further level of integration brings under the management of LIS applications the control of robotic instrumentation through worklists – large yet simple text files that contain the required pipetting commands for the execution of a routine – written in language supported by the robot's control software (Leclair *et al.*, 2004b; Leclair and Scholl, 2005). Pipetting and sample flow management logic can be integrated to transform isolated instruments into components of fully integrated processing platforms. If the intuitive approaches of trained forensic analysts can be understood in sufficient detail to permit their description in a rules-based system, then the system can dynamically alter default processing schemes when certain pre- or mid-process conditions are met. This type of customization can include every aspect of automated pipetting, such as changing source and / or destination containers (i.e. cherry-picking), altering transfer volumes and changing liquid pipetting specifications. Sample flow management logic can manage sample re-processing queues to regroup samples that share the same point of reintroduction in the processing scheme or similar modified pipetting schemes, and build batches optimized for pipetting efficiency. An integrated capability promotes higher first-pass processing success by reducing wasteful analytical attempts under non-ideal conditions, which may prove mission-critical for many scarce casework samples. Such a system provides flexibility with the processing platform that exceeds what may normally be achieved manually.

Convicted offender databanking was the first forensic process to employ extensive automation. These specimens are collected under controlled conditions similar to those employed for clinical genetic testing, making for ideal specimens for genetic analysis. As most of the wet chemistry is similar from one forensic application to the next, the experience from automated processing of large numbers of convicted offender specimens has been leveraged into other areas of forensic processing. However, variation in sample input attributes greatly increases the processing contingencies that must be handled and thereby increases the design complexity for an automated processing system. In that respect, casework and MFI samples, being collected at crime / disaster scenes, often present compounded problems linked to substantial variation in substrate, cell / tissue type and quality / quantity of recoverable biological material, which calls for chemistries and robotic pipetting schemes supporting a larger range of quality / quantity of input material. The increase in precision and reproducibility afforded by automated pipetting devices provides a consistency that may improve the quality of STR data generated from compromised samples.

Casework samples also present additional intricacies at the evidence screening step as many samples need to be localized and cut away from larger pieces of evidence, all samples need to be assessed to confirm human origin of the recovered material, the body fluid involved, their suitability for ensuing DNA extraction and genetic analysis, and the amount of material to be processed in order to meet processing platform sample input range specifications. A substantial, largely manual front-end processing step is necessary to qualify the samples for further processing and direct them to an appropriate wet chemistry processing protocol. Throughput improvements for this front-end step may come from improved stain visualization technologies, streamlined presumptive tests and clerical technologies (i.e. LIS-supported voice recording, speech-to-text software, digital photography and tactile computer screens) to facilitate and expedite accurate information capture.

In summary, substantial improvements in throughput can be readily realized through the implementation of LIS-supported automation of pre-data analysis steps. Development activities should focus on enhancing efficiencies through aids to evidence screening, and on the extended integration of automated liquid handlers with LIS systems. Continued progress in these fields is essential if reductions in casework backlogs are to parallel those for convicted offenders.

11.4   Analysis of STR data

The review of STR data can be relatively straightforward for pristine, single-source specimens such as those collected from convicted offenders. Still, regardless of sample type, a number of samples will present some anomalous features that may interfere with the final allele call and genotype assignment. Most anomalies (e.g. pull-up, saturated and split peaks, elevated baseline, heterozygote ratio imbalance, elevated stutter, profile slope) will present electrophoretic data signatures that can be recognized by suitably designed algorithms and quality metrics. In fact, automated quantitative assessment of quality metrics is likely to be more efficient at enforcing quality control thresholds than through human review. However, not all qualitative problem scenarios can be anticipated and coded into quality control algorithms intended for automated data review systems. In order to ensure that all anomalous results are scrutinized, the experience of trained analysts must bear on the final interpretation of the data. Historically, it has been customary for all data to be subjected to dual review by separate analysts, although the practice can become a limiting throughput factor in certain environments.

Many large data banks have developed automated data review packages, or 'expert systems', integrated in their LIS systems, and several software packages are commercially available as well. Quantitative measurements used to affirm quality control thresholds form the basis of these review applications. Detection of the presence of mixtures, genuine or due to cross-contamination within a processing batch, are additional features available in current applications. Electrophoretic data signatures of frequently encountered anomalies, commonly referred to as 'rules', are used to filter data sets and flag electropherograms that require a second, human review. The available published data (Kadash *et al.*,

2004; *http://www.cstl.nist.gov/biotech/strbase/pub_pres/NIST_FSSi3_Mar2006. pdf*) on the performance of these systems on convicted offender type samples indicate that, on average, 30% of allele calls are flagged by at least one rule. This significant number of flagged allele calls may reflect the removal of subjectivity of human review from the process, resulting in more consistent examination of every locus, as well as the enforcement of more stringent thresholds to ensure the capture of all problematic data. However, very high allele call concordance (>99.9% in most studies) between manual and automated review results was reported, which provides support to the feasibility of replacing a two-person review process with a single reviewer assisted by a computer algorithm.

Casework data review presents additional complexities linked to the often-compromised nature of the casework specimens. Increased slopes across profiles, dropped-out alleles and mixtures are regularly encountered with these specimens. An automated data review system configured for databanking needs may indicate which casework samples need further review.

Analysis of mixtures

Some of the most complex crime scene evidence samples from a data analysis standpoint contain biological material from multiple contributors, typically encountered with sexual assault evidence. The simplest and most common mixture scenario stems from a contamination of the differential lysis sperm fraction with the epithelial fraction originating from the victim. These are the easiest analytical circumstances since the known genotype of the victim may be subtracted from the mixed profile to establish a list of obligate alleles for the perpetrator. If the male contributor to the mixture represents the major profile in the mixture, it may prove possible to deduce a complete male genotype, which is ideally suited for a search of both the convicted offender and crime scene indices of data banks.

The next most common mixture situation involves a sperm fraction holding two male profiles – the perpetrator's and the victim's consensual partner – reflecting consensual intercourse in the hours / days preceding the assault. In the much rarer instance of collective sexual assaults, multiple male contributors may be recovered from the sperm fraction, generating a much more complex STR profile that may prove impossible to deconvolute with technologies available at this time. Under these last two scenarios, no procedure akin to differential lysis can alter the major:minor contributor ratio to facilitate discrimination between contributors.

When the investigation has produced individual(s) suspected to have contributed to a mixture, a first investigative step is to attempt to exclude the suspect(s) as contributor(s) to the mixture. In the absence of suspects, the perpetrators' genotypes must be dissected out from the mixed profile to allow for a search against a convicted offender data bank. The deconvolution of a mixture involves three steps: the ascertainment of the number of contributors, the estimation of the proportion of the individual contributions to the mixture and the establishment of a list of possible contributing genotypes that could explain the mixed profile along with probability estimates. When the ratio of components of a two-contributor mixture is more than $1:3$, peak height or peak area information may allow an analyst to visually resolve the major and minor components of the mixture. With other ratios of contributors, the ascertainment of the number of contributors can prove a challenge in itself, and an incorrect assessment may have dramatic effects on the interpretation of testing results (Paoletti *et al.*, 2005). Many approaches to mixture deconvolution have been proposed over the years (Weir *et al.*, 1997; Clayton *et al.*, 1998; Evett *et al.*, 1998a, 1998b; Gill *et al.*, 1998; Curran *et al.*, 1999; Perlin and Szabady, 2001; Fung and Hu, 2002; Wang *et al.*, 2002; Cowell 2003; Mortera *et al.*, 2003; Bill *et al.*, 2005; Curran *et al.*, 2005; Cowell *et al.*, 2006), but a consensus on mixture interpretation guidelines has yet to emerge (Gill *et al.*, 2006; Schneider *et al.*, 2006).

Several LIS systems designed to assist with mixture deconvolution are being evaluated in the community. Bill *et al.* (2005) proposed a computerized algorithm to estimate the proportion of the individual contributions in two-person mixtures and to rank the genotype combinations based on minimizing a residual sum of squares, eliminating unreasonable genotypic combinations. Perlin and Szabady (2001) and Wang *et al.* (2002) have proposed *linear mixture analysis* and *least square deconvolution* models, respectively, to estimate mixture proportion and enumerate a complete set of possible genotypes that may explain the mixed profiles. Cowell *et al.* (2006) have proposed a model unifying, under a single Bayesian network model, many of the elements of the above-mentioned models. No model currently takes into account all potential technical complications such as dropped-out alleles, stutter and excessive profile slopes.

11.5   Bioinformatics

The comparison of genotypes is an intrinsic component of all forensic genetic analysis. Comparative analysis can link crime scene to suspect, victim to relative and reveal cases that share perpetrators. Most computations performed in data banks involve searches for perfect or partial genetic matches between a query and an entry in one of the data bank indices. System innovations currently focus on 'familial searching' algorithms (Bieber and Lazer, 2004; Bieber, 2006; Bieber *et al.*, 2006). As nearly half of jailed inmates in the USA are reported to have at least one close relative who has been incarcerated (Bieber *et al.*, 2006), the use of likelihood ratio computations to detect potential parent–child or sibling relationships between an unknown perpetrator's genotype recovered in crime scene evidence and that of a catalogued offender offers new investigative

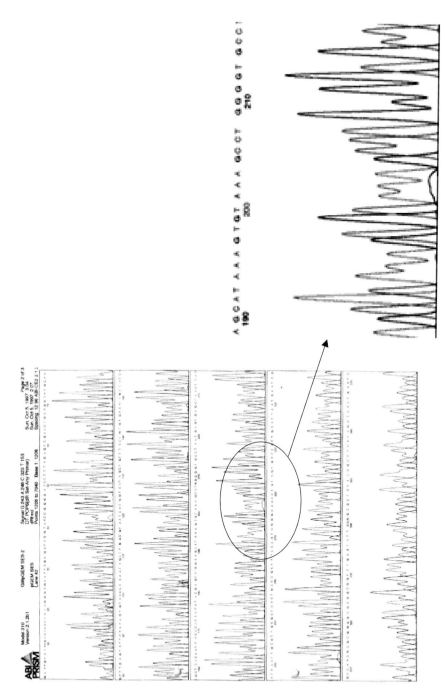

Plate 2.1 ABI Prism sequencing readout illustrating the capacity of sequence determination by automated fluorescence sequencing. The expanded sequence panel indicates the variety of peak heights from which unambiguous base calls can be made

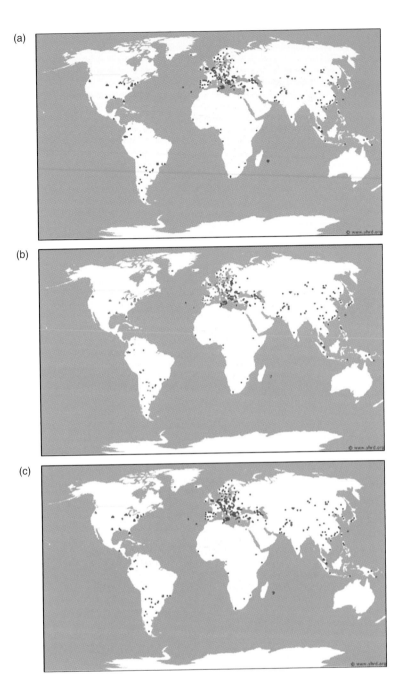

Plate 9.1 Global distributions of three Y-STR haplotypes consisting of DYS19, DYS389I, DYS389II, DYS390, DYS391, DYS392, DYS393 and DYS385a-b according to the Y Chromosome Haplotype Reference Database (YHRD, Release '19' from August 2006, *http://www.yhrd.org/index.html*). Dots represent analysed population samples included in the database, with red dots indicating the presence of the respective haplotype (and dark red dots indicating a higher frequency than light red dots) and blue dots indicating the absence of the haplotype. (a) The most characteristic haplotype of the Eastern European metapopulation: 17,13,30,25,10,11,13,10–14. (b) The most characteristic haplotype of the Western European metapopulation: 14,13,29,24,11,13,13,11–14. (c) The most frequent haplotype in U.S. African Americans: 15,13,31,21,10,11,13,16,17. Note that population samples are depicted according to their contemporary place of sampling, e.g. samples from the Americas belong to populations of Native American/European or Asian/African genetic ancestry. For additional information, see text

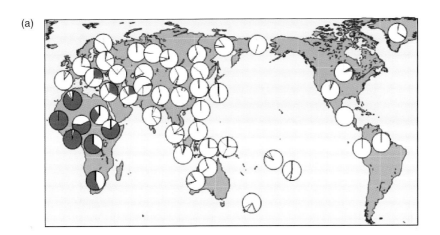

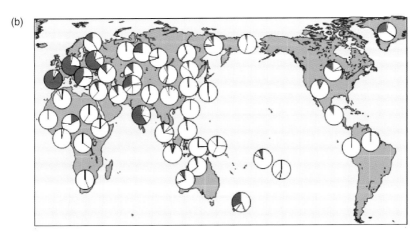

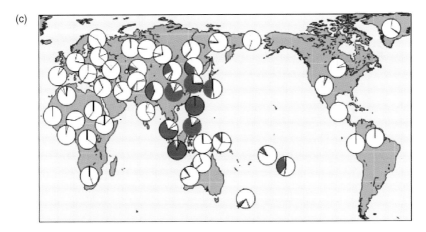

Plate 9.2 Global distribution and frequency estimates of three Y-SNP haplogroups. Each pie chart diagram represents a single population sample: (a) haplogroup E with predominantly African distribution; (b) haplogroup R1 with predominantly European distribution; (c) haplogroup O with predominantly East Asian distribution. Note that only population samples authentic to the respective geographical regions (at least concerning prehistoric times) are depicted, e.g. samples from the Americas all represent Native American populations. The figure was prepared from a compilation of published data made by Dr Chris Tyler-Smith (Hinxton, UK). For additional information, see text

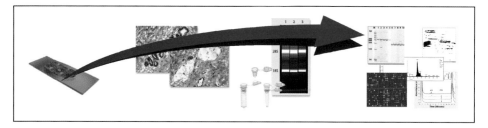

Plate 10.1 A schematic representation of forensic laboratory activity from laser microdissection of a biological sample to single tandem repeat typing of nuclear DNA extracted from its cells

Plate 10.3 High-magnification images of the LCM process, showing the polymer 'tearing' the target cell

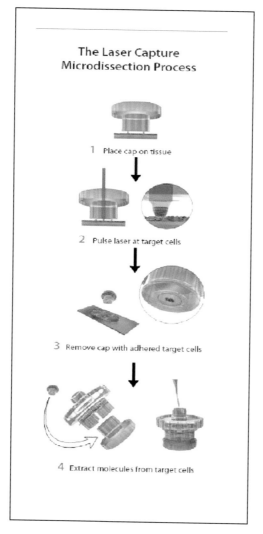

Plate 10.2 Diagram summarizing the laser capture microdissection process described in the text

COLOUR PLATE

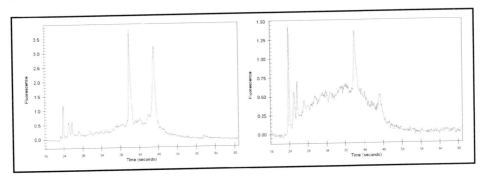

Plate 10.4 Comparison between the micro-isolation activity of the laser capture microdissection (on the left) and another laser microdissecting device (on the right) on the quality of RNA recovery

Plate 10.5 Representation of the microdissection system recovering, by gravity, the isolated cell sample in the down-standing collection tube cap

Plate 10.6 High-magnification images of the laser catapulting system device, showing the isolation of the target sample through a non-harmful laser pulse towards the up-standing collection tube cap

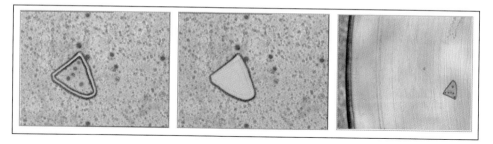

Plate 10.7 Laser microdissection of a small group of cells smeared on a glass. The sequence shows the different phases of the activity: *cutting* and *collection* of the target sample, which will undergo nuclear DNA extraction and STR typing

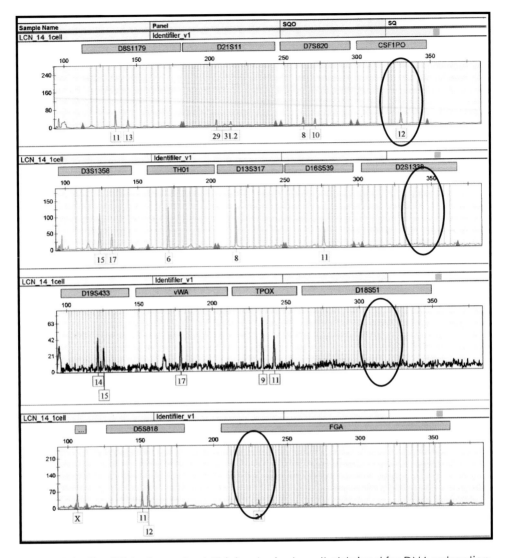

Plate 10.8   The STR typing output. This is a typical result obtained for DNA extraction from a very small number of human diploid cells. Allele drop-outs are shown within the circles as a consequence of the very small amount of template DNA. For details, see the text

possibilities, and could lead to a substantial increase in cold hit rates, if implemented.

Over the last decade, the forensic community was confronted with several challenges, unique in scope of work and technical complexity, in the wake of large-scale MFIs. From airliner mishaps to the World Trade Center attacks to tsunamis, large increases in the number of victims and extreme fragmentation and/or decay of recovered remains have led to a considerable paradigm shift in the way that identification initiatives are conducted and in the laboratory infrastructure needed to handle such events. DNA typing has taken a major role in large-scale events involving high body fragmentation. The complexity of these initiatives is such that bioinformatics have become a crucial component for identification mandates to be met (Brenner and Weir, 2003; Cash *et al.*, 2003; Leclair, 2004; Leclair *et al.*, 2004a; Biesecker *et al.*, 2005; Budowle, 2005; Leclair *et al.*, 2007).

All MFIs are unique in the circumstances of the incident, and the required identification solutions will vary between incidents. There are common limitations to these events that create unique demands on LIS applications to support DNA-based identifications. The first limitation pertains to unavailability of a reference genotype, usually generated from trace biological material recovered from a personal effect (i.e. personal hygiene items), for a proportion of the victims of MFIs as probative personal effects are often destroyed in many events (e.g. air crashes). This limitation makes it necessary to supplement direct matching algorithms with computationally-intensive large-scale kinship analysis and parentage trio searching routines. The second limitation is linked to the obligation of performing parentage trio searching routines, and results from the inaccuracy of some reported biological relationships as a consequence of incorrect information capture. Traditional triangulation methods, by which the third member of a parentage trio can be quickly located within a genotype data set with the help of a list of obligate alleles derived from the two surviving members of the trio, rely on the accuracy of next-of-kin self-reported biological relationships. An alternate procedure is required to detect parentage trios solely on a genetic basis, immune to sample accessory information errors. This is especially important for missing persons databasing (MPD) initiatives as, contrary to MFIs, MPDs are open-ended, long-term initiatives and, as such, problematic data may go undetected indefinitely. The alternative procedure is also crucial for events where families are among the victims (e.g. most airliner mishaps) as pedigrees must be re-assembled, solely on a genetic basis, from within the victims' genotype data set (Leclair *et al.*, 2004a). This alternative procedure calls for the evaluation of every possible parentage trio involving each victim and every pair of related and unrelated next-of-kin, an evaluation that may amount to a very substantial and escalating computing workload.

Many additional event-related limitations add more complexity still. More often than not in MFIs, remains recovery is partial, a proportion of recovered remains has incurred significant thermal / chemical / bacterial decay leading to the production of partial genotypes during analysis, many potential contributors to parentage trios involving older victims may be pre-deceased and many victims may have few next-of-kin who can be used as genotypic references. Computing solutions are expected to provide ways to mitigate the lack of complete STR profiles from the recovered remains and the absence of important reference genotypes.

Finally, the scale of the incident has significant impact on computing parameters. As pair-wise comparisons are the mainstay of computing efforts in MFI victim identification, the computing load increases nearly exponentially with the number of victims. These complicating factors add considerable complexity to the design of bioinformatics tools required to produce the necessary identification inferences.

As much as the World Trade Center appears to have been the most demanding identification initiative to date, the circumstances of the incident could have substantially increased the complexity of DNA-based identifications. The data processing contingencies would have been very different if the collapse of the World Trade Center towers had trapped the normal weekday occupancy of 50000 instead of 2749, or had included families (Leclair *et al.*, 2007). High body fragmentation / high remains dispersal incidents involving 100000 or 1000000 casualties as a result of natural, accidental or terrorist activity are within the realm of possibilities. Bioinformatic tools that can handle this sample load have been developed, however it is unclear whether the STR loci currently in use in forensics would provide sufficient discrimination power to support large-scale kinship analysis and parentage trio searching algorithms. These scenarios can be simulated and conclusions drawn as to the genetic marker set and computing capabilities that would be required in varying circumstances.

11.6   Conclusion

The scale and complexity of current forensic projects, such as large backlogs of data bank / casework samples, large-scale MFIs or MPDs, have reached a level where these projects can greatly benefit from the utilization of LIS-assisted automation technologies. The throughput and sample tracking capabilities afforded by LIS-driven automated liquid handlers will contribute to substantial reductions in backlogs and improvements in process quality, sample tracking confidence and turn-around time. Although these instruments have tremendous capacity, they address aspects of laboratory processing that tend not to limit throughput. The labor-intensive procedures of data review, data collation, enforcement of quality control standards and formatting results for submission to databases stand to benefit from the introduction of automated data review applications. Expert systems to assist with casework mixture deconvolution are emerging and will benefit from a consensus on mixture interpretation from the forensic community. Familial searching algorithms would provide investigative assistance in cases where perpetrators are not yet registered in offender data banks. The victim identification of large-scale MFIs and MPDs has greatly benefited from LIS-based large-scale comparative genotyping applications to achieve their mandates, and larger throughput capabilities are being developed. These LIS-based technologies will continue to evolve and take full advantage of the information content of available forensic genotypic data.

11.7 References

Bieber, F.R. (2006) Turning base hits into earned runs: improving the effectiveness of forensic DNA data bank programs. *J. Law Med. Ethics* **34**: 222–233.

Bieber, F.R., Brenner, C.H. and Lazer, D. (2006) Human genetics. Finding criminals through DNA of their relatives. *Science* **312**: 1315–1316.

Bieber, F.R. and Lazer, D. (2004) Guilt by association: should the law be able to use one person's DNA to carryout surveillance on their family? Not without a public debate. *New Scientist* **184**: 20.

Biesecker, L.G., Bailey-Wilson, J.E., Ballantyne, J., Baum, H., Bieber, F.R., Brenner, C., *et al.* (2005) Epidemiology. DNA identifications after the 9/11 World Trade Center attack. *Science* **310**: 1122–1123.

Bill, M., Gill, P., Curran, J., Clayton, T., Pinchin, R., Healy, M. and Buckleton, J. (2005) PENDULUM – a guideline-based approach to the interpretation of STR mixtures. *Forensic Sci. Int.* **148**: 181–189.

Brenner, C.H. and Weir, B.S. (2003) Issues and strategies in the DNA identification of World Trade Center victims. *Theor. Popul. Biol.* **63**: 173–178.

Budowle, B., Bieber, F.R. and Eisenberg, A.J. (2005) Forensic aspects of mass disasters: strategic considerations for DNA-based human identification. *Leg. Med. (Tokyo)* **7**: 2302–2343.

Cash, H.D., Hoyle, J.W. and Sutton, A.J. (2003) Development under extreme conditions: forensic bioinformatics in the wake of the World Trade Center disaster. *Pac. Symp. Biocomput.* **8**: 638–653.

Clayton, T.M., Whitaker, J.P., Sparkes, R. and Gill, P. (1998) Analysis and interpretation of mixed forensic stains using DNA STR profiling. *Forensic Sci. Int.* **91**: 55–70.

Cowell, R.G. (2003) FINEX: a Probabilistic Expert System for forensic identification. *Forensic Sci. Int.* **134**: 196–206.

Cowell, R.G., Lauritzen, S.L. and Mortera, J. (2006) Identification and separation of DNA mixtures using peak area information. *Forensic Sci. Int.* [Epub ahead of print] doi: 10.1016/j.forsciink.2006.03.021.

Curran, J.M., Gill, P. and Bill, M.R. (2005) Interpretation of repeat measurement DNA evidence allowing for multiple contributors and population substructure. *Forensic Sci. Int.* **148**: 47–53.

Curran, J.M., Triggs, C.M., Buckleton, J. and Weir, B.S. (1999) Interpreting DNA mixtures in structured populations. *J. Forensic Sci.* **44**: 987–995.

Evett, I.W., Foreman, L.A., Lambert, J.A. and Emes, A. (1998a) Using a tree diagram to interpret a mixed DNA profile. *J. Forensic Sci.* **43**: 472–476.

Evett, I.W., Gill, P.D. and Lambert, J.A. (1998b) Taking account of peak areas when interpreting mixed DNA profiles. *J. Forensic Sci.* **43**: 62–69.

Frégeau, C.J., Leclair, B., Bowen, K.L. and Fourney, R.M. (2003) The National DNA Data Bank of Canada – A laboratory bench retrospective on the first year of operation. *Int. Congr. Ser.* **1239**: 621–625.

Fung, W.K. and Hu, Y.Q. (2002) The statistical evaluation of DNA mixtures with contributors from different ethnic groups. *Int. J. Legal Med.* **116**: 79–86.

Gill, P. (2001) An assessment of the utility of single nucleotide polymorphisms (SNPs) for forensic purposes. *Int. J. Legal Med.* **114**: 204–210.

Gill, P., Brenner, C.H., Buckleton, J.S., Carracedo, A., Krawczak, M., Mayr, W.R., Morling, N., Prinz, M., Schneider, P.M. and Weir, B.S. (2006) DNA commission of the International Society of Forensic Genetics: Recommendations on the interpretation of mixtures. *Forensic Sci. Int.* [Epub ahead of print] doi: 10.1016/j.forsciink. 2006.04.016.

Gill, P., Sparkes, R., Pinchin, R., Clayton, T., Whitaker, J. and Buckleton, J.S. (1998) Interpreting simple STR mixtures using allele peak areas. *Forensic Sci. Int.* **91**: 41–53.

Gill, P., Werrett, D.J., Budowle, B. and Guerrieri, R. (2004) An assessment of whether SNPs will replace STRs in national DNA databases – joint considerations of the DNA working group of the European Network of Forensic Science Institutes (ENFSI) and the Scientific Working Group on DNA Analysis Methods (SWGDAM). *Sci. Just.* **44**: 51–53.

Hagelberg, E., Gray, I.C. and Jeffreys, A.J. (1991) Identification of the skeletal remains of a murder victim by DNA analysis. *Nature* **352**: 427–429.

Hall, A. and Ballantyne, J. (2003) Novel Y-STR typing strategies reveal the genetic profile of the semen donor in extended interval post-coital cervicovaginal samples. *Forensic Sci. Int.* **136**: 58–72.

Jobling, M.A. and Gill, P. (2004) Encoded evidence: DNA in forensic analysis. *Nat. Rev. Genet.* **5**: 739–751.

Kadash, K., Kozlowski, B.E., Biega, L.A. and Duceman, B.W. (2004) Validation study of the TrueAllele automated data review system. *J. Forensic Sci.* **49**: 660–667.

Kimpton, C.P., Gill, P., Walton, A., Urquhart, A., Millican, E.S. and Adams, M. (1993) Automated DNA profiling employing multiplex amplification of short tandem repeat loci. *PCR Methods Appl.* **3**: 13–22.

Leclair, B. (2004) Large-scale comparative genotyping and kinship analysis: evolution in its use for human identification in mass fatality incidents and missing persons databasing. In: *Progress in Forensic Genetics*, vol. 10 (C. Doutremépuich and N. Morling, eds), Elsevier Science, Amsterdam, The Netherlands, pp. 42–44.

Leclair, B., Frégeau, C.J., Bowen, K.L. and Fourney, R.M. (2004a) Enhanced kinship analysis and STR-based DNA typing for human identification in Mass Fatality Incidents: the Swissair Flight 111 disaster. *J. Forensic Sci.* **49**: 939–953.

Leclair, B. and Scholl, T. (2005) Application of automation and information systems to forensic genetic specimen processing. *Expert Rev. Mol. Diagn.* **5**: 241–250.

Leclair, B., Schwensen, C. and Kupferschmid, T.D. (2004b) Casework sample processing and automation: demystifying robotics and tracking systems. *Proceedings of 103rd Semi-annual Seminar of the California Association of Criminalists*, Foster City, CA.

Leclair, B., Shaler, R., Carmody, G.R., Eliason, K., Hendrickson, B.C., Judkins, T., Norton, M.J., Sears, C. and Scholl, T. (2007) Bioinformatics and human identification in mass fatality incidents: the World Trade Center disaster, *J. Forensic Sci.* in press.

Lygo, J.E., Johnson, P.E., Holdaway, D.J., Woodroffe, S., Whitaker, J.P., Clayton, T.M., Kimpton, C.P. and Gill, P. (1994) The validation of short tandem repeat (STR) loci for use in forensic casework. *Int. J. Legal Med.* **107**: 77–89.

Mortera, J., Dawid, A.P. and Lauritzen, S.L. (2003) Probabilistic expert systems for DNA mixture profiling. *Theor. Popul. Biol.* **63**: 191–205.

Paoletti, D.R., Doom, T.E., Krane, C.M., Raymer, M.L. and Krane, D.E. (2005) Empirical analysis of the STR profiles resulting from conceptual mixtures. *J. Forensic Sci.* **50**: 1361–1366.

Perlin, M.W. and Szabady, B. (2001) Linear mixture analysis: a mathematical approach to resolving mixed DNA samples. *J. Forensic Sci.* **46**: 1372–1378.

Schneider, P.M., Gill, P. and Carracedo, A. (2006) Editorial on the recommendations of the DNA commission of the ISFG on the interpretation of mixtures. *Forensic Sci. Int.* **160**, 89.

Scholl, T. (2004) The development of highly integrated and automated forensic specimen processing platforms. *Proceedings of the Cambridge Health Institute's 6th Biannual DNA Forensics Conference*, McLean, VA.

Scholl, T., Pyne, M.T. and Ward, B.E. (1998) High throughput automated forensic genetic analysis. *Proceedings of the Ninth International Symposium on Human Identification*, Scottsdale, AZ.

Schoske, R., Vallone, P.M., Kline, M.C., Redman, J.W. and Butler, J.M. (2004) High-throughput Y-STR typing of U.S. populations with 27 regions of the Y chromosome using two multiplex PCR assays. *Forensic Sci. Int.* **139**: 107–121.

Sullivan, K.M., Hopgood, R. and Gill, P. (1992) Identification of human remains by amplification and automated sequencing of mitochondrial DNA. *Int. J. Legal Med.* **105**: 83–86.

Urquhart, A., Kimpton, C.P., Downes, T.J. and Gill, P. (1994) Variation in short tandem repeat sequences – a survey of twelve microsatellite loci for use as forensic identification markers. *Int. J. Legal Med.* **107**: 13–20.

Wang, T., Xue, N. and Wickenheiser, R. (2002) Least square deconvolution (LSD): a new way of resolving STR/DNA mixture samples. *Proceedings of the 13th International Symposium on Human Identification*, 7–10 October, Phoenix, AZ.

Weir, B.S., Triggs, C.M., Starling, L., Stowell, L.I., Walsh, K.A. and Buckleton, J. (1997) Interpreting DNA mixtures. *J. Forensic Sci.* **42**: 213–222.

12

Statistical presentation of forensic data

Mark A. Best

12.1 Introduction

Genetic information may be found in bone, teeth, skin and other soft tissue, tears, sweat, saliva, hair roots, earwax, semen, vaginal fluid, urine and blood. Examples would include saliva on a cigarette butt or drinking glass, or skin cells on a steering wheel or glass. About 95% of human nuclear DNA is non-coding DNA, the so-called 'junk' DNA. These non-coding regions are the areas that are examined in DNA forensic testing (Varsha, 2006). DNA forensic fingerprinting began in Leicester, UK, when in 1984 Alec Jeffreys discovered hypervariable loci made up of approximately 10 to 1000 tandemly repeated sequences, each typically 10–100 base pairs in length (Gill *et al.*, 1985; Jeffreys *et al.*, 1985a, 1985b). Applications of molecular DNA analysis include criminal, immigration and civil cases. Murder, assault, rape, paternity testing and family and body identification are all issues that can be addressed with these techniques.

12.2 Techniques

Though the typing of restriction fragment length polymorphisms (RFLPs) was originally the primary method of DNA forensic analysis, the main forensic DNA analysis method of today uses autosomal short tandem repeat (STR) loci, which are usually typed by multiplex polymerase chain reaction (PCR) methods. The profiles from an STR multiplex analysis are displayed in a computer-generated

Molecular Forensics.   Edited by Ralph Rapley and David Whitehouse
Copyright 2007 by John Wiley & Sons, Ltd.

graph called an electropherogram. The sources of ambiguity of an electrophero-gram profile are mixtures of samples, degradation of DNA, allelic drop-out, spurious peaks (technical artefacts) or false peaks.

In selected cases, such as degraded samples or when very small amounts of the sample are available for study, 'mini' STR markers that use shorter ampli-cons (Biesecker *et al.*, 2005), autosomal single nucleotide polymorphisms (SNPs) or markers on the mitochondrial DNA or on the Y chromosome are more useful (Jobling and Gill, 2004). The two largest DNA databases in the world, the UK National DNA Database (>3.4 million profiles) and the USA DNA database (>3.5 million profiles), use multiplex autosomal STR methods. Because SNPs have much lower heterozygosities than STRs, data from a larger number of SNP loci would be required to obtain the same discrimination potential. There is currently a 21-locus autosomal SNP multiplex that has been introduced (Dixon *et al.*, 2005).

Differential lysis

In rape cases there may be a mixing of the victim's DNA and the rapist's DNA in the vaginal vault or other anatomical locations. Differential lysis is a tech-nique used for vaginal fluid/semen mixtures that concentrates the sperm by lysing the victim's epithelial cells. This process thus eliminates the masking of the rapist's DNA by the victim's DNA. Though first described in 1985 (Jeffreys *et al.*, 1985b), the original protocol is still used today.

Autosomes (nuclear DNA)

Currently, the most common method for forensic analysis of nuclear DNA is to use a multiplex system for STRs (see Chapter 5). There are about 20 000 known STR loci, however about 20 select loci are used in most forensic cases (DNA Advisory Board, 2000). The STRs that are commonly used were selected because they are independent: that is, the inheritance of one locus does not influence the inheritance of another locus. The 13 core loci used by the US FBI CODIS system are: TPOX, D3S1358, FGA, D5S818, CSF1PO, D7S820, D8S1179, TH01, VWA, D13S317, D18S51, D21S11 and D16S539 (Budowle *et al.*, 1998) Along with these 13 core loci, the amelogenin locus for sex deter-mination is also examined (Sullivan *et al.*, 1993). For each locus, a genotype for the sample will be identified; then the frequency of such a genotype can be calculated using population genotypes. An example of this process is presented later in this chapter, using the loci in Table 12.1.

At times the SNP approach may be utilized for nuclear DNA analysis, instead of the more popular (and discriminating) STRs, because SNPs are more useful in analysing degraded DNA than are STRs.

Table 12.1 The CODIS loci used for DNA profiling

Locus	TPOX	D3S1358	FGA	D5S818	CSF1PO	D7S820	D8S1179
Genotype Frequency							
Locus	THO1	vWA	D13S317	D16S539	D18S51	D21S11	AMEL
Genotype Frequency							

Y Chromosome

The Y chromosome is useful to trace relationships among males and to evaluate multiple males in a rape case and in male–female mixtures (see Chapter 9). There are at least 246 Y-chromosome SNP markers (Y-SNPs) and at least 227 Y-chromosome STR markers (Y-STRs) described (Butler, 2003). Another author reports that there are at least 219 STR sites on the Y chromosome (Kayser *et al.*, 2004). The STR approach is the usual method utilized for DNA analysis of the Y chromosome.

Mitochondrial DNA (mtDNA)

Mitochondrial DNA is more durable and plentiful than nuclear DNA (see Chapter 8). However, it is less discriminating than nuclear DNA because it is transmitted only from mother to children, and has less variation between individuals than nuclear DNA. It is usually used when the nuclear DNA is degraded, such as when only hair, bone or teeth are present. There is a report of finding suitable mtDNA in guts of maggots (fly larvae) that have fed on human tissue and been collected in death investigations (Wells *et al.*, 2001). There are many more copies of mtDNA than there are copies of nuclear DNA. There are usually 200–1700 copies of mtDNA per cell (Holland and Parsons, 1999). Heteroplasmy is a special attribute of mtDNA. Heteroplasmy is when a person has two or more different mtDNA sequences present. Human mitochondria have 16 569 base pairs that comprise 37 genes. A region called the discrimination loop, or 'D-loop' is a non-coding control region that has a fair amount of variation, and so is used for DNA testing of SNPs (Butler, 2001). The mtDNA D-loop region has two hypervariable areas called HV1 and HV2. The HV2 area has a region of length heteroplasmy identified as the homopolymeric cytosine stretch (C-stretch).

Investigations of the mtDNA coding region using an SNP method to increase forensic discrimination has been recently described (Coble *et al.*, 2006). Statistical analysis of mtDNA is typically a 'counting method'. Thus instead of calculating the match probability, which will typically be in the range of 0.005–0.025 (Budowle *et al.*, 1999), a match is based on how many times a specific sequence is found in a population database (Monson *et al.*, 2002; Parson *et al.*, 2004).

Guidelines for reporting mtDNA analysis (Scientific Working Group on DNA Analysis Methods, SWGDAM, 2003) list three options as outcomes. These are:

1. *Exclusion* – the known and unknown samples have two or more nucleotide differences, thus the samples can be excluded as being from the same person or maternal lineage.

2. *Inconclusive* – one nucleotide difference between unknown sample and the known sample.

3. *Cannot Exclude* – sequences from known and unknown samples have a common base at each location or a common length variant in the HV2 C-stretch, thus the samples cannot be excluded as being from the same person or maternal lineage.

Messenger ribonucleic acid (RNA)

Messenger RNA (mRNA) appears to have a role in the identification of body fluids such as blood, saliva and semen (Juusola and Ballantyne, 2003). This approach may be utilized more in the future as a replacement for serological and protein analysis.

12.3   Laboratory issues

Laboratories that run forensic tests should adhere to high-quality standards and be accredited. The laboratories should engage in proficiency testing and the results from such tests made available for review. A positive control, a negative control and a reagent blank sample should be run for all tests. Inadvertent transfer of DNA, chain of custody and contamination are major areas of concern for DNA forensic evidence and could contribute to undesirable variation in a case.

Whenever possible, samples should be divided into two or more parts so that additional tests or repeat tests can be performed. If possible, any additional tests should be run by different personnel from those conducting the first test, and in a different laboratory if possible. The best opportunity for justice, for a wrongly implicated innocent person, is an independent re-test (National Research Council, 1996).

12.4 Statistical analysis

Only about 0.1% (about 3 million bases) of a person's DNA differs from one person compared to another person. However, the amount of genetic material that is shared depends on the degree of relatedness that one person is to another person (see Table 12.2) (Wenk *et al.*, 1996; Jobling *et al.*, 1997; Tzeng *et al.*, 2000; Wenk and Chiafari, 2000). Thus relatives share more genetic material or loci than non-relatives. Statistical methods for estimating the probability of a close relative matching the suspect's DNA profile are discussed in the literature (Li and Sacks, 1954; National Research Council, 1996).

Using the 13 CODIS loci as an example, there is about a 10% chance that one of the 13 sites will match in two individuals. Two close relatives, such as a parent, child or sibling, will have about four or five sites that match out of the 13 loci. However, for people who are not close relatives there is less than one in a trillion that all 13 sites will match.

How are these probabilities calculated? The 10% per single site is based on population genetics. For example, if a Hispanic individual had a 10% match to the Hispanic population at each site, the probability that all 13 sites match another Hispanic individual would be:

$$0.1 \times 0.1 \times 0.1 \times 0.1 \times 0.1 \times 0.1 \times 0.1 \times 0.1 \times 0.1 \times 0.1 \times 0.1 \times 0.1 \times 0.1$$
$$= 0.1^{13}$$
$$= \text{less than one in a trillion}$$

The match probability (P_m) (or random match probability) is obtained by using the *product rule*. Multiplying the probability for each site by the probabilities for all the other sites is called the product rule. How do we calculate the probability that two or more people have the same genetic profile? Here is the process in five steps.

Table 12.2 Comparison of relationship and genes

Degree of relatedness	Identical	First-degree relative	Second-degree relative	Third-degree relative
Genes in common	100%	50%	25%	12.5%
Relationship	Identical twin	Parent Child Full sibling	Grandparent Grandchild Half-sibling Aunt/uncle Niece/nephew	Great grandparent Great grandchild Great aunt/uncle Great or half- niece/-nephew Half-aunt/-uncle First cousins

Step 1

Identify the STRs for the individual genotypes. An example of individual geno-types for the CODIS loci is presented in Table 12.3.

Table 12.3 Profile of a hypothetical individual set of genotypes for the CODIS loci

Locus	TPOX	D3S1358	FGA	D5S818	CSF1PO	D7S820	D8S1179
Genotype Frequency	8, 8	15, 15	24, 25	10, 13	11, 11	11, 10	12, 13
Locus	THO1	vWA	D13S317	D16S539	D18S51	D21S11	AMEL
Genotype Frequency	9, 9.3	14, 16	10, 11	11, 11	12, 13	29, 31	XY

Step 2

Identify what proportion of the population is at each allele being tested. The population database is created or a pre-existing database identified. The product rule method is based on the assumption that the population shows Hardy-Weinberg equilibrium. This means that the population has random mating and thus the allele selections are statistically independent from a common gene pool, so the results are independent associations. This assumption is based on the Hardy-Weinberg Principle. The Hardy-Weinberg Principle is an elementary formula for population genetics. A chi-squared analysis can be used to determine if the population is in Hardy-Weinberg equilibrium. This test for independence cannot prove independence, but it can find dependence if it exists.

Thus, as you see in Table 12.4, the three possible genotype frequencies in the offspring are:

Table 12.4 Punnett square for Hardy-Weinberg equilibrium for alleles 'A' & 'a' at a given locus

		Female	
		A (p)	a (q)
M a l e	A (p)	AA (p^2)	Aa (pq)
	a (q)	Aa (pq)	aa (q^2)

f(AA) = p^2
f(Aa) = 2pq
f(aa) = q^2

Therefore the equation for genotype frequencies is:

P^2 + 2pq + q^2 = 1

There are several DNA databases currently, with the two largest being the UK National DNA Database (>3.4 million profiles) and the US FBI CODIS (Combined DNA Index System) database (>3.5 million profiles). The CODIS system tests for 13 STRs and the amelogenin sex test. Recently a 16-loci multiplex system has been introduced as a possible upgraded system with more loci. (Greenspoon *et al.*, 2004) The CODIS system includes at least four population substructure reference databases.

Step 3

Calculate the frequency for each locus. For a homozygous genotype:

P = p^2

For the data in Table 12.3, the genotype (15, 15) at locus D3S1358 is calculated as follows. From the population reference database, the genotype 15 frequency is 17.3%, therefore:

P = p^2 = 17.3% × 17.3% = 0.173 × 0.173 = 0.173^2 = 0.030 = 3.0%

For a heterozygous genotype:

P = 2pq

For the data in Table 12.3, the genotype (14, 16) at locus vWA is calculated as follows. From the population reference database, genotype 14 is estimated to be 15.7% and genotype 16 is 22.7%, therefore:

P = 2pq = 2(15.7%)(22.7%) = 2(0.157) (0.227) = 0.071 = 7.1%

The two step 3 frequency estimates can be seen in Table 12.5 and the same process would be undertaken for each locus.

By calculating all of the loci, a profile such as the one in Table 12.6 is obtained.

Table 12.5 Profile with population frequencies estimated for two loci

Locus	TPOX	D3S1358	FGA	D5S818	CSF1PO	D7S820	D8S1179
Genotype Frequency	8, 8	15, 15 3.0%	24, 25	10, 13	11, 11	11, 10	12, 13
Locus	THO1	vWA	D13S317	D16S539	D18S51	D21S11	AMEL
Genotype Frequency	9, 9.3	14, 16 7.1%	10, 11	11, 11	12, 13	29, 31	XY

Table 12.6 Example of a hypothetical complete 13-loci profile (plus AMEL locus)

Locus	TPOX	D3S1358	FGA	D5S818	CSF1PO	D7S820	D8S1179
Genotype Frequency	8, 8 3.4%	15, 15 3.0%	24, 25 4.2%	11, 13 12.8%	11, 11 7.5%	10, 10 6.9%	13, 14 8.2%
Locus	THO1	vWA	D13S317	D16S539	D18S51	D21S11	AMEL
Genotype Frequency	9, 9.3 10.1%	14, 16 7.1%	11, 11 1.4%	11, 12 4.7%	16, 18 8.4%	29, 30 12.3%	XY Male

Step 4

Calculate the DNA profile probability for the multilocus genotype using the product rule. The product rule (multiplication rule) is:

$$\text{Probability of random match} = P_m = (P_1)(P_2)(P_3) \ldots (P_n)$$

Using the data from Table 12.6 will result in the probability:

$$P_m = (0.034)(0.030)(0.042)(0.128)(0.075)(0.069)(0.082)(0.101)(0.071)$$
$$(0.014)(0.047)(0.084) \ (0.123) = 9.395^{-19}$$

Thus the probability that two people (other than identical twins) have the profile in Table 12.6 is less than one in a hundred trillion.

Step 5

Calculate confidence limits for each allele. An upper confidence limit should be calculated for each allele frequency in the population. This gives the confidence that the profile is unique, given the population of N unrelated people. The upper 95% confidence limit (95% UCL) has the following formula:

$$P + 1.96\sqrt{P(1-P)/N}$$

where P is the observed frequency and N is the number of chromosomes studied.

The lower 95% confidence limit (95% LCL) has the following formula:

$$P - 1.96\sqrt{P(1-P)/N}$$

12.5 Other issues

'Ceiling' principle

In cases where the sample and suspect belong to a subpopulation then a 'ceiling' should be placed on the estimate of the profile. This will be a conservative correction of the estimate. It has been used to compensate for any undetected subpopulation that may exist in the population database.

The 95% UCL should be used or 0.10, whichever is larger.

The 95% LCL should be used or 0.05, whichever is smaller.

If a subpopulation (population substructure) database is used, instead of the entire population database, then the ceiling principle would not need to be considered.

The prosecutor's fallacy and defence fallacy

An example of the 'prosecutor's fallacy' is given next. Making a statement like 'There is only a one-in-a-trillion chance that the defendant is innocent' is a statement about guilt or innocence, and is not true. The true statement is 'There is a one-in-a-trillion chance that the forensic sample came from an individual other than the defendant'.

An example of the 'defence fallacy' is as follows. Suppose that a murder occurred in a city with a population of 5 million people. A match was found between the suspect (defendant) and a stain sample from the crime scene. The match probability was calculated to be one in a million. In a city of 5 million, about five people would have a matching profile. Thus the defence argues that the odds are 5 to 1 that the defendant is innocent. This assumes that each of the five people have an equal probability of guilt. This would only be true if the DNA evidence was the only evidence and was used in isolation of any other facts pertaining to the case.

Bayes' theorem

Bayes' theorem may be utilized in select circumstances, but it is not a commonly accepted statistical method for presenting forensic evidence. This method is based on prior probabilities based on certain facts of a specific case. The major argument against using Bayes' theorem is that the prior probabilities may be subjective.

Likelihood ratio

A commonly accepted way of expressing the likelihood of matching evidence is by calculating the likelihood ratio (LR). Here is an example of a criminal case utilizing the LR method:

$$LR = \frac{P(\text{evidence originated from suspect})}{P(\text{evidence originated from an unrelated person in the population})}$$

$$LR = \frac{\text{Probability that the prosecutor is correct}}{\text{Probability that the defence is correct}}$$

$$= \frac{P(\text{prosecetor's hypothesis})}{P(\text{defence hypothesis})}$$

The likelihood that two people are siblings can be calculated like this:

$$LR = \frac{P[\text{allele(s) would match if two people were siblings}]}{P[\text{allele(s) would match if two people were unrelated}]}$$

12.6   Special situations

DNA mixtures

A simple mixture of two individuals may be evaluated by comparing sizes of the fluorescent peaks of the electropherogram (Clayton *et al.*, 1998). Otherwise, the likelihood ratio can be used (Weir *et al.*, 1997; Evett *et al.*, 1991), calculations based on the fluorescent peak area (Evett *et al.*, 1998) or evaluating PCR stutter (Gill *et al.*, 1998; Ladd *et al.*, 2001).

Complex settings

Special considerations need to be observed in sex-change (sex-reversal) individuals. In this case the AMEL genotype would be in disagreement with the phenotype. Also, in individuals who have had a bone marrow transplant (BMT), there would be atypical results, such as a DNA mixture.

12.7 References

Biesecker, L.G., Bailey-Wilson, J.E., Ballantyne, J., Baum, H., Bieber, F.R., Brenner, C., *et al.* (2005) DNA identification after the 9/11 World Trade Center Attack. *Science* **310**: 1122–1123.

Budowle, B., Moretti, T.R., Niezgoda, S.J. and Brown, B.L. (1998) *Proceedings of the Second European Symposium on Human Identification*, Promega Corporation, Madison, WI, pp. 73–88.

Budowle, B., Wilson, M.R., Di Zinno, J.A., Stanffer, C., Fasano, M.A., Holland, M.M. and Monson K.L. (1999) Mitochondrial DNA regions HVI and HVII population data. *Forensic Sci. Int.* **103**: 23–35.

Butler, J.M. (2001) Additional DNA markers. *Forensic DNA Typing*, Academic Press, New York, p. 121.

Butler, J.M. (2003) Recent developments in Y-short tandem repeat and Y-single nucleotide polymorphism analysis. *Forensic Sci. Rev.* **15**: 91–111.

Clayton, T.M., Whitaker, J.P., Sparkes, R.L. and Gill, P. (1998) Analysis and interpretation of mixed forensic stains using DNA STR profiling. *Forensic Sci. Int.* **91**: 55–70.

Coble, M.D., Vallone, P.M., Just, R.S., Diegoli, T.M., Smith, B.C. and Parsons, T.J. (2006) Effective strategies for forensic analysis in the mitochondrial DNA coding region. *Int. J. Legal Med.* **120**: 27–32.

Dixon, L.A., Murray, C.M., Archer, E.J., Dobbins, A.F., Koumbi, P. and Gill, P. (2005) Validation of a 21-locus autosomal SNP multiplex for forensic identification purposes. *Forensic Sci. Int.* **154**: 62–77.

DNA Advisory Board (2000) Statistical and population genetics issues affecting the evaluation of the frequency of occurrence of DNA profiles calculated from pertinent population database(s). *Forensic Sci. Commun.* **2**(3) (accessed 1/10/06 at www.fbi. gov/hq/lab/fsc/backissu/july2000/dnastat.htm).

Evett, I.W., Buffery, C., Willot, G. and Stoney, D.A. (1991) A guide to interpreting single locus profiles of DNA mixtures in forensic cases. *Forensic Sci. Int.* **31**: 41–47.

Evett, I.W., Gill, P. and Lambert, J.A. (1998) Taking account of peak areas when interpreting mixed DNA profiles. *Forensic Sci. Int.* **43**: 62–69.

Gill, P., Jeffreys, A.J. and Werrett, D.J. (1985) Forensic application of DNA 'fingerprints'. *Nature* **318**: 577–579.

Gill, P., Sparkes, R. and Buckleton, J.S. (1998) Interpretation of simple mixtures when artifacts such as stutters are present – with special references to multiplex STRs used by the Forensic Science Service. *Forensic Sci. Int.* **95**: 213–224.

Greenspoon, S.A., Ban, J.D., Pablo, L., Grouse, C.A., Kist, F.G., Tomsey, C.S., *et al.* (2004) Validation and implementation of the PowerPlex 16 BIO System STR multiplex for forensic casework. *J. Forensic Sci.* **49**: 71–80.

Holland, M.M. and Parsons, T.J. (1999) Mitochondrial DNA sequence analysis – validation and use for forensic casework. *Forensic Sci. Rev.* **11**: 21–49.

Jeffreys, A.J., Wilson, V. and Thein, S.L. (1985a) Hypervariable 'minisatellite' regions in human DNA. *Nature* **314**: 67–73.

Jeffreys, A.J., Wilson, V. and Thein, S.L. (1985b) Individual-specific 'fingerprints' of human DNA. *Nature* **316**: 76–79.

Jobling, M.A. and Gill, P. (2004) Encoded evidence: DNA in forensic analysis. *Nat. Rev.* **5**: 739–751.

Jobling, M.A., Pandya, A. and Tyler-Smith, C. (1997) The Y chromosome in forensic analysis and paternity testing. *Int. J. Legal Med.* **110**: 118–124.

Juusola, J. and Ballantyne, J. (2003) Messenger RNA profiling: a prototype method to supplant conventional methods for body fluid identification. *Forensic Sci. Int.* **135**: 85–96.

Kayser, M., Kittler, R., Giler, A., Hedman, M., Lee, A.C., Mohyuddin, A., *et al.* (2004) A comprehensive survey of human Y chromosomal microsatellite. *Am. J. Hum. Genet.* **74**: 1183–1197.

Ladd, C., Lee, H.C., Yang, N. and Bieber, F.R. (2001) Interpretation of Complex Forensic DNA Mixtures. *Croat. Med. J.* **42**: 244–246.

Li, C.C. and Sacks, L. (1954) The derivation of joint distribution and correlation between relatives by the use of stochastic matrices. *Biometrics* **10**: 347–360.

Monson, K.L., Miller, K.W.P., Wilson, M.R., Di Zinno, J.A. and Budowle, B. (2002) The mtDNA population database: an integrated software and database resource of forensic comparison. *Forensic Sci. Comm.* **4** (accessed at www.fbi.gov/hq/lab/fsc/backissu/april2002/miller1.htm.

National Research Council Committee on DNA Forensic Science (1996) *An Update: The Evaluation of Forensic DNA Evidence.* National Academy Press, Washington, DC.

Parson, W., Brandstatter, A., Alonso, A., Brandt, N., Brinkmann, B., Carracedo, A., *et al.* (2004) The EDNAP mitochrondrial DNA population database (EMPOP) collaborative exercises: organization, results and perspectives. *Forensic Sci. Int.* **139**: 215–226.

Scientific Working Group on DNA Analysis Methods (SWGDAM) (2003) Mitochondrial DNA (mtDNA) nucleotide sequence interpretation. *Forensic Sci. Commun.* **5**(2).

Sullivan, K.M., Mannucci, A., Kimpton, C.P. and Gill, P. (1993) A rapid and quantitative DNA sex test: fluorescence-based PCR analysis of X–Y homologous gene amelogenin. *Biotechniques* **15**: 636–641.

Tzeng, C.H., Lyou, J.Y., Chen, Y.R., Hu, H.Y., Lin, J.S., Wang, S.Y. and Lee, J.C. (2000) Determination of sibship by PCR-amplified short tandem repeat analysis in Taiwan. *Transfusion* **40**: 840.

Varsha (2006) DNA fingerprinting in the criminal justice system: an overview. *DNA Cell Biol.* **25**: 181–188.

Weir, B.S., Triggs, C.M., Starling, L., Stowell, L.I., Walsh, K.A.J. and Buckleton, J. (1997) Interpreting DNA mixtures. *Forensic Sci. Int.* **42**: 213–222.

Wells, J.D., Introna Jr., F., Di Vella, G., Campobasso, C.P., Hayes, J. and Spelling F.A.H. (2001) Human and insect mitochondrial dna analysis from maggots. *J. Forensic Sci.* **46**: 685–687.

Wenk, R.E. and Chiafari, F.A. (2000) Distinguishing full siblings from half-siblings in limited pedigrees. *Transfusion* **40**: 44–47.

Wenk, R.E., Traver, M. and Chiafari, F.A. (1996) Determination of sibship in any two persons. *Transfusion* **36**: 259–262.

13

Protein profiling for forensic and biometric applications

Mikhail Soloviev, Julian Bailes, Nina Salata and Paul Finch

13.1   Introduction

This chapter evaluates the usability of known protein markers for quantitative protein profiling assays applicable to forensic and biometric applications. We will discuss the use of competitive displacement assays and peptidomics technology in the design of multiplex protein assays for laboratory-based and potentially scene-of-crime applications. Competitive displacement assays embody the most accurate multiplex protein affinity assay system to date, applicable to a wide range of protein profiling applications. The peptidomics technology is the most generic and multiplatform-compatible affinity assay system to date, applicable for protein identification, quantification and expression profiling. It relies on proteolytic protein digestion, requires only the availability of small peptide fragment(s) and is therefore capable of a reliable analysis of denatured, partially degraded proteins and protein fragments. Protein microarrays and peptidomics are combined in a single system capable of simple yet quantitative protein analysis from real samples (i.e. imperfectly stored, partially degraded samples, scene-of-crime applications, etc.).

13.2   Protein assays in molecular forensics: current status

Biometric technologies aim to identify individuals using biological traits, the major one being based on fingerprint recognition (~50% of current biometrics market share), followed by face, hand and iris recognition (12%, 11% and 9%, respectively), and voice and signature recognition (6% and 2%, respectively). Biologically-based technologies in forensic sciences are mostly limited to polymerase chain reaction (PCR)-based DNA 'fingerprinting'. Whilst DNA can be matched against databases relatively easily, in the absence of a match, tissue samples, body fluids or stains may still reveal vital intelligence information useful for fitting the person's biometric or behavioural profiles or for gathering additional forensic information. DNA assays are most suitable for matching and identification of individuals, whilst protein and metabolite profiles are more representative of the state of health, lifestyle, behavioural patterns, sample origin (tissue/organ/time), the severity and type of trauma or the cause of death. DNA profiling is incapable of monitoring these. mRNA profiling may be, but mRNAs are restricted to their respective tissues and cells (and are very unstable molecules), whilst proteins and metabolites are often secreted and can be detected/measured in body fluids, e.g. blood/serum/urine/saliva. Single protein assays have been proposed for forensic analysis before, but so far the research has been fragmented and until very recently no technical capabilities existed for highly parallel and quantitative analysis of proteins and scene-of-crime assays.

Affinity immunoassays have been widely used for achieving specific and sensitive detection of analytes of interest. A wide range of existing assay types includes colorimetric (Khosravi *et al.*, 1995; Garden and Strachan, 2001; Moorthy *et al.*, 2004), radiometric (Ahlstedt *et al.*, 1976; Yalow, 1980; Raja *et al.*, 1988), fluorescence (Kronick, 1986; Dickson *et al.*, 1995) and chemiluminescence (Rongen *et al.*, 1994) detection methods, each incorporating their own form of labels, such as radioisotopes for radiometric tests or organic dyes for fluorescence-based assays. The performance of these tests is usually restricted to central laboratories because of the need for long assay times, complex and expensive equipment and highly trained individuals, but there is an ever-increasing market for new, faster, more accurate and cost-effective diagnostics that can be supplied in kit format for use in the field. The ability to multiplex such assays is also highly desirable, allowing for the simultaneous detection of more than one analyte in a given assay.

Liquid chromatography assays

Size exclusion chromatography permits the separation of molecules by physical size and thus can be harnessed for use in a potentially very attractive immunoassay format. The assay can be in the form of a large-scale gel filtration set-up utilizing a column packed with gel media of chosen fractionation range (e.g. Sephadex®), or a small – scale set-up by way of MicroSpin® (both from GE Healthcare) columns. Gel filtration media separate molecules according to size, with larger molecules being eluted first, followed by smaller molecules in order of their size. This can be harnessed in an immunoassay format by exploiting the size differences between antigen–antibody complexes and unbound antibody and antigen. Antibody molecules are ~150 kDa; other proteins are of variable size but far larger than peptides, which are ~1–3 kDa, and small-molecule drugs (<1 kDa). Therefore a column containing gel with a fractionation range capable of resolving molecules of ~150 kDa (antibody) from molecules of larger size (antigen–antibody complexes) would be suitable for use in such an assay format. The assay is compatible with both competitive and non-competitive formats, both of which can be designed a number of different ways:

1. **Competitive:**

   a. Fluorescently labelled reference competes for antibodies with sample antigen. Fluorescence intensity is inversely proportional to sample antigen concentration.

   b. Fluorescently labelled sample competes for antibodies with unlabelled reference. Fluorescence intensity is proportional to sample antigen concentration.

2. **Non-competitive:**

   a. Fluorescently labelled antibodies are mixed with sample to form labelled antibody–antigen complexes. Fluorescence intensity is proportional to sample antigen concentration.

   b. Fluorescently labelled sample is mixed with antibodies to form labelled antibody–antigen complexes. Fluorescence intensity is proportional to sample antigen concentration.

Given the small size of peptides and small drug molecules, labelling antibodies may pose problems in that the column may not be capable of resolving the relatively small differences in size between unbound antibody and antibody–antigen complexes. In this instance labelled antibody would provide a potentially misleading result as to the amount of sample antigen concentration and an overall low level of detection. A better choice would be to label antigen (either sample or reference), which will be dramatically different in size to antibody within the system, and thus if the column was to elute some free antibody with antibody–antigen complexes then this would not provide such a misleading result.

Eluted buffer should be collected from the column up until the point where only antibody–antigen complexes have had a chance to elute. This collection can then be scanned with a spectrofluorimeter. Standard curves should be made for a range of known sample antigen concentrations by using single antigen–antibody pairs. Following successful completion of this, a multiplex assay may be attempted, which can be achieved though using a range of fluorescent tags for different analytes. For multiplexing, competitive assays with fluorescently labelled reference are the only suitable option because they provide the simplest way to assign different fluorescent tags to specific analytes. Miniaturizing the assay results in a format more applicable to the field-based studies. A microspin format (e.g. similar to Bio-Rad Micro Bio-Spin columns) or small cartridge-like gravity-flow format (e.g. similar to NAP columns from GE Healthcare) with columns containing the same gel capable of quantitative separation of antigen–antibody complexes from unbound antibody presents such a solution. A portable spectrofluorimeter could then be used to check for fluorescence intensity as described above, making the format applicable for scene-of-crime applications.

Immunochromatography assay format

Otherwise known as lateral flow or strip tests, immunochromatographic assays exhibit a number of highly desirable benefits, including a user-friendly format, rapid sample turnaround, as well as being relatively inexpensive to produce. Such features make them ideal for affordable field-based or point-of-care testing. Pregnancy tests are an example of an immunochromatographic assay designed for qualitative determination of human chorionic gonadotrophin (hCG) in urine for early detection of pregnancy. Lateral flow devices have been applied successfully to a wide range of detection applications, including aflatoxin B_1 in pig feed (Delmulle *et al.*, 2005), botulism neurotoxins in foods (Sharma *et al.*, 2005) and drugs of abuse (Niedbala *et al.*, 2001). In their simplest form, lateral flow tests consist of a porous membrane strip such as nitrocellulose that has a band of capture antibodies immobilized at a discrete point across its width (Qian and Bau, 2004). This mixture diffuses through the membrane towards the capture line where hybridization occurs, detectable by standard fluorescence detection methods. A control line is added to the membrane after the capture line, consisting of immobilized antibodies that bind to the reporter but not to the analyte of interest. Lateral flow assays are compatible with competitive and non-competitive immunoassay formats, outlined below:

1. **Competitive:**

   a. Fluorescently labelled reference competes with sample antigen for antibodies discretely spotted on a membrane strip. Fluorescence intensity is inversely proportional to sample antigen concentration.

   b. Fluorescently labelled sample competes with unlabelled reference for antibodies discretely spotted on a membrane strip. Fluorescence intensity is directly proportional to sample antigen concentration.

c. Fluorescently labelled antibody is added to sample and then run on membrane strip with reference antigen discretely spotted at a certain point. Fluorescence intensity is directly proportional to sample antigen concentration.

2. Non-competitive:

a. Fluorescently labelled sample antigen binds to antibodies at a discrete capture line on membrane strip. Fluorescence intensity is directly proportional to sample antigen concentration.

b. Fluorescently labelled antibody binds to sample antigen spotted at a discrete capture line on a membrane strip. Fluorescence intensity is proportional to sample antigen concentration.

In practice neither of the non-competitive forms of the assay are ideal because they require the sample to be either fluorescently labelled or spotted onto the membrane strip prior to testing. Minimum sample preparation should be the focus and for this reason the first and third competitive formats are the most appealing. Competitive forms of lateral flow tests can also be used for small-molecule analytes with single antigenic determinants that are incompatible with sandwich forms of the assay (Qian and Bau, 2004). Predictive tools have been developed by computer modelling, allowing simulation and optimization of a device and reducing the number of laboratory experiments needed in the development of lateral flow devices (Qian and Bau, 2003, 2004).

The membrane strip can be scanned, with intensity of fluorescence at the capture line quantitatively indicating the presence or absence of sample antigen. The assay may first be assessed using single antigen–antibody pairs for which standard curves can be plotted for a range of known sample concentrations, followed by a multiplexed approach once conditions for each analyte have been optimized. Multiplexing may be achieved by spatially separating the capture lines for each analyte or by using a single capture line and assigning a fluorescent tag of different colour to each analyte of interest.

Immunochromatographic assays are highly suited to incorporation into kit format that would usually consist of a mould essentially enabling the user to plug-and-play by simply adding a sample to a defined region of the membrane. Such kits offer standardization of use each time for position/sample application/detection, and can be achieved with minimal effort.

Other formats

Quartz crystal microbalance. The quartz crystal microbalance (QCM) is a simple and convenient method of quantitatively measuring very small masses in real time. It is a form of acoustic wave technology, so called because an acoustic wave is the mechanism of detection. The velocity or amplitude of the wave can be changed as it passes through the surface of the material, and such changes can be detected by measuring the frequency or phase characteristics of the sensor. Any changes can be correlated to physical interactions occurring on the surface of the sensor, such as binding of sample analyte to surface-immobilized antibody, which would result in a frequency decrease due to a mass increase from the biological interactions (Thompson *et al.*, 1986; Muratsugu *et al.*, 1993; Sakai *et al.*, 1995). Such devices are classified by the mode of wave that propagates through or on the substrate, and, of the many wave modes available, shear-horizontal surface acoustic wave (SH-SAW) sensors are best as biosensors due to their superior ability to operate with liquids (Drafts, 2001). A special class of these is the Love wave sensor, which consists of a series of coatings on the surface of the device, including a final coating with biorecognition capability. The Love wave sensor has demonstrated excellent sensitivity (Gizeli *et al.*, 1992, 1993; Kovacs and Venema, 1992; Du *et al.*, 1996) and the ability to detect anti-goat IgG in solution in the concentration range of $3 \times 10^{-8} - 10^{-6}$ M (Gizeli *et al.*, 1997). The QCM technology has been applied to a number of other fields, such as detection of a class A drug (Attili and Suleiman, 1996) and mutations in DNA (Su *et al.*, 2004), and is commercially available from a number of providers, e.g. Attana Sensor Technologies Ltd [*www.attana.com*] or Akubio Ltd [*www.akubio.com*]. Figure 13.1g summarizes the basic principles behind the technology.

Rupture event scanning. The QCM technology is used in another form of biosensor known as rupture event scanning (REVS). However, rather than being used to measure mass increase, as is the case with other QCM-based detection systems such as the Love wave sensor, a piezoelectric substrate is used to detect the binding and estimate the affinity of analyte binding to antibodies covalently attached to the surface by detecting acoustic noise produced from the rupturing of bonds between antigens and antibodies. By applying an alternating voltage to gold electrodes on the upper and lower surfaces of a disc of crystalline quartz, and monotonously increasing the voltage and thus the amplitude of the transverse oscillation of the QCM, Cooper *et al.* (2001) demonstrated a novel way of directly, sensitively and quantitatively detecting virus particles bound to specific antibodies immobilized on the QCM surface. Figure 13.1h depicts the general principles involved in REVS. Both the Love wave sensor and REVS are suitable for forensic (field and laboratory) applications and have the additional advantage of providing label-free detection of molecules, allowing interactions to be monitored between unmodified reactants.

Surface plasmon resonance and BIAcore. Another label-free approach to assaying an analyte of interest in a sample is by way of the BIAcore system. The BIAcore system is based on surface plasmon resonance (SPR), an optical phe-

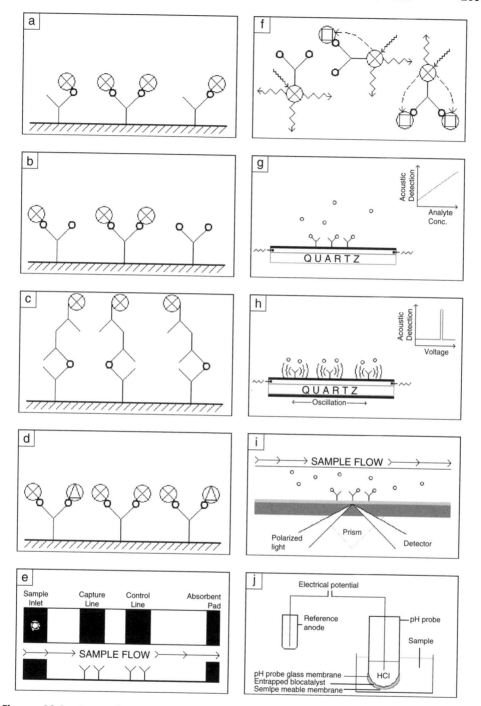

Figure 13.1 Assay formats: (a) direct; (b) competitive; (c) sandwich; (d) two-colour; (e) lateral flow; (f) quenching/FRET; (g) QCM; (h) REVS; (i) BIAcore/SPR; (j) potentiometric

nomenon occurring when polarized light is reflected off a thin metal film under conditions of total internal reflection (Kretschmann, 1971). In the BIAcore system, this thin metal film composed of gold forms the floor of a small flow cell, and can be modified so that antibodies are immobilized on its surface. Running buffer is passed continuously through the flow cell and a sample containing the analyte of interest can be injected into this mobile phase. Any interactions that occur between sample antigen and immobilized antibody results in a change in the local refractive index that subsequently changes the SPR angle (Leonard et al., 2003). The change in intensity of reflected light is plotted against time, producing a sensorgram (Merwe, 2003). BIAcore systems have been used for a range of detection applications, including detection of cancer biomarkers in human saliva (Yang et al., 2005). The underlying principles of this technology are presented in Figure 13.1i.

Amperometric, potentiometric, capacitance and ion-selective field-effect transistor-based sensors. Electrochemical sensors are capable of detecting changes in a solution's electrochemical properties that result from binding or biocatalytical events (D'Orazio, 2003). Electrochemical biosensors are the most common form used for clinical analysis and amperometry is the technique usually applied (D'Orazio, 2003). Amperometric devices exploit electroreactive substances and were demonstrated first in 1962 for the detection of glucose in blood (Clark and Lyons, 1962), which spawned continual progression in their application to this field (Wang, 2001). Potentiometric sensors monitor the electrical potential difference between a reference electrode and an indicator electrode that is placed in the sample solution. The potential difference is related to the concentration of analyte in solution in a logarithmic manner (Leonard et al., 2003). The indicator electrode, although immersed in the sample solution, is surrounded by a semi-permeable membrane coated with an entrapped biocatalyst (see Fig 13.1j). Changes in capacitance were used in the recent development of a novel formaldehyde-sensitive biosensor (Ben Ali et al., 2006), further expanding the range of electrochemical-based detection techniques. Finally, ion-selective field-effect transistor (ISFET)-based biosensors were recently separated as a distinct class of electrochemical sensors, and have a promising scope for application (Dzyadevych et al., 2006).

13.3 Novel technologies and the remaining challenges

Microarray- and macroarray-based protein assays

Nucleic acid amplification using PCR has revolutionized forensic science by providing a powerful tool for mitochondrial and chromosomal DNA typing and other nucleic acid-based analyses. Hundreds of publications available so far report the use of PCR amplification in forensics and related disciplines. DNA microarrays are another relatively new technology that allows a much higher multiplexity of nucleic acid analysis (by hybridizing a probe simultaneously to many thousands of spots of various cDNAs or oligonucleotides in a single array). Today DNA microarrays have become a routine tool in transcriptomics, but have not so far been used widely in forensic science, where PCR remains the preferred tool (due to its sensitivity). One of the difficulties is that DNA analysis using microarrays requires orders of magnitude larger amounts of initial material to be used compared to PCR-based analysis, therefore if only little material is available, which is often the case in forensic applications, microarrays may be unsuitable for the job. For example, PCR sensitivity is ultimately one DNA molecule, whilst microarrays (i.e. hybridization analysis) would typically require several hundred nanograms of mRNA. Protein microarrays would seem to be an obvious successor to DNA arrays and many formats have already been attempted. Unlike nucleic acids, however, protein targets are typically non-homogeneous and affinity capture agents are often poorly characterized, making the experiments difficult to perfect and reproduce. Moreover, running multiple affinity assays in parallel (multiplexing) is not possible due to the heterogeneity of antibody affinities to their protein targets.

Unlike genomic DNA, which is present at one of two copies per cell, proteins are often present in millions of copies per single cell. This could make a protein affinity assay (which is analogous to DNA hybridization) possible in cases where DNA microarrays may fail. There is very little, if any, research published so far on the use of protein microarrays in forensics, and this is not surprising taking into account the technical difficulties of working with immobilizing proteins in their affinity-active conformational states.

Protein array-based proteomics has many advantages over traditional proteomic techniques based on two-dimensional gels and chromatography (Soloviev and Terrett, 2005). Highly multiplexed assays can be achieved by spatially separating antibodies on the membrane when spotting, and standard curves can be plotted for a range of known sample antigen concentrations, from which unknown sample concentrations can be derived. An experiment with a single protein chip can supply information on thousands of proteins simultaneously and this provides a considerable increase in throughput. A micro- or macroarray is compatible with both competitive and non-competitive immunoassay formats:

1. **Competitive:**

   a. Fluorescently labelled reference competes with sample antigen for antibodies discretely spotted on solid support. Fluorescence intensity is inversely proportional to sample antigen concentration.

   b. Fluorescently labelled antibody is added to sample and then hybridized with reference antigen discretely spotted on solid support. Fluorescence intensity is directly proportional to sample antigen concentration.

2. Non-competitive:

  a. Fluorescently labelled sample binds to antibody discretely spotted on solid support. Fluorescence intensity is directly proportional to sample antigen concentration.

  b. Fluorescently labelled antibody (primary or secondary) binds to sample antigen discretely spotted on solid support. Fluorescence intensity is proportional to sample antigen concentration.

A novel modification of the competitive displacement strategy has been reported recently (Barry *et al.*, 2003) and is a technique that can utilize almost any antibody or indeed other types of affinity reagents or their mixtures (affinity heterogeneity is not an issue) and allows a high degree of multiplexing (the number of proteins assayed being limited only by antibody availability). It is tolerant towards high levels of non-specific binding and does not require any potentially interaction-disrupting labelling of the experimental samples. It is capable of quantitative comparison of unlabelled experimental samples over a wide concentration range. Other advantages of the method include its relative simplicity and low cost (only a single labelled reference sample per series and a single array per sample are required), and intrinsic signal normalization and compatibility with known signal amplification techniques (e.g. ELISA, electrochemical luminescence, RCAT, DNA fusions, etc.); see Soloviev *et al.*, 2004, and Barry and Soloviev, 2004, for a more detailed review and additional references).

Affinity and combinatorial peptidomics approaches

Affinity peptidomics is the most generic affinity assay system reported to date, which is applicable for protein identification and quantification for forensic and biometric applications. It is multiplatform compatible (liquid chromatography, microarrays, microfluidics and mass spectrometry). In the peptidomic approach the assayed mixture of proteins (typically a mixture of heterogeneous proteins) is enzymatically digested (e.g. with trypsin) prior to affinity capture to form a homogeneous mixture of short peptides. These peptides can also be predicted by *in silico* digestion of individual proteins or protein databases. Capture agents can therefore be specifically designed for all or a subset of suitable peptides from each of the proteins. Such peptides are fully predictable on the basis of protein sequence alone (or even predicted sequence). The use of mass spectrometry (e.g. MALDI-ToF-MS) for a direct confirmation of the identity of the species captured provides an additional advantage compared to the more usual method of detection in which fluorescently labelled captured species are scanned to give a spatially resolved image of the array.

In peptidomics, each protein is broken down into many smaller components, resulting in the availability of a large range of peptides with less heterogenic physical and chemical properties (which are also more predictable). Large numbers of proteolytic peptides allow *multiple independent assays* for the same protein target to be performed, thus also increasing the reliability of the assay. Peptidomics enables a high-throughput screening of proteins (e.g. in a microarray format) and has several advantages over the affinity capture of intact proteins:

1. As peptides are much more stable and robust than proteins, protein denaturation and degradation is not an issue since only one or a few intact peptides would be required for the analysis. Affinity peptidomics does not have to struggle with unstable or degraded proteins, it uses proteolytically digested samples and relies on anti-peptide affinity reagents (e.g. antibodies, but these can also be antibody mimics (Nord *et al.*, 1997), molecularly imprinted reagents (Haupt and Mosbach, 1998), etc.

2. Peptides are also particularly suited for detection by mass spectrometric techniques, such as MALDI-ToF-MS, for direct analysis of samples on a solid substrate such as microarrays. The peptide mass range is such that isotopic resolution is easily achieved and hence fully quantitative analysis is possible (e.g. using isotopically labelled standards as in AQUA (Gerber *et al.*, 2003), MCAT (Cagney and Emili, 2002) or ICAT (Gygi *et al.*, 1999) approaches.

3. Digestion of cellular fractions or even intact tissues results in the release of peptides, which in most cases will contain more than one hydrophilic peptide (Soloviev and Finch, 2005) per protein, thus improving the assay.

4. Antibody can be against linear unfolded fragments, not native folded proteins, and therefore peptide 'antigens' can be more easily generated, such as by chemical synthesis of *in silico* predicted peptides (as opposed to traditionally used fully folded proteins or their fragments) (Soloviev *et al.*, 2003).

5. Such affinity reagents can be obtained at lower costs and in a truly high-throughput manner and against most antigenic peptides, and their specificities and affinities can be more easily controlled.

6. The affinity peptidomics approach is suitable for both microarray-based assays (for quick/routine applications, whether field or laboratory-based) and analytical mass spectrometry-based analysis (e.g. quantitative mass spectrometry using AQUA / ICAT / MCAT approaches), suitable for resolving difficult cases or independent confirmation of the array data if required.

Another approach suitable in principle for use in forensic applications is the combinatorial approach to peptidomics analysis (Soloviev and Finch, 2005). It utilizes the original peptidomics approach where protein samples are proteolytically digested using one or a combination of proteases, but in place of affinity purification the peptide pool is depleted through selective chemical binding of a subset of peptides to a solid support. This combinatorial approach utilizes the selective chemical reactivities of the side chains of individual amino acids, and thus is the nearest to a sequence-dependent analysis (e.g. DNA-based analyses). Together, the affinity peptidomics (fast and high throughput) and the combinatorial approaches (more sequence-dependent analysis) provide a viable alternative to traditional protein analysis techniques (protein preservation–separation pathway). Peptidomics approaches provide the most generic protein assay system to date, applicable for protein identification, quantification and expression profiling. These are multiplatform compatible and are capable of the analysis of partially degraded proteins, which makes them especially suitable for forensic applications. Combining the peptidomics approach with a protein microarray platform will eventually yield a new miniature tool for on-site analysis and scene-of-crime applications.

13.4 Protein markers for use in forensic and biometric applications

A survey of over 15 years' worth of relevant forensic and biomedical literature reveals a shortlist of candidate protein markers for use in protein-based forensic analysis. The selection (currently based on over 200 publications, not listed here due to size limitation) is summarized in Table 13.1. For clarity we have divided these into five main categories: *Biometrics, Blood Origin, Lifestyle, Time of Death, Trauma and Death.* We have also included a few 'small molecule' metabolites (e.g. testosterone, etc.) as these are also relevant to molecular forensics *and* antibodies are commercially available *and* the same assay format (CDA; see Barry *et al.*, 2003; Barry and Soloviev, 2004) can be used. We have recently obtained 40 antisera against 20 of the ~100 targets listed in Table 13.1. Prediction of antigenic peptide sequences is a crucial part of any antibody generation programme; the chosen markers and their respective anti-peptide antibodies are shown in Table 13.2 (for abbreviated protein names, see the legend to Table 13.1). The choice of the peptides was made on the basis of their predicted immunogenicity using in-house software (manuscript in preparation).

Another parameter requiring special attention is the choice of the solid support for the microarrays. Extensive data on protein array production parameters have been published in recent years by us and others (Kuznezow *et al.*, 2003; Soloviev and Terret, 2005). Protein arrays have been produced on plastic or glass slides (most similar to DNA microarrays), hydrogels, filters or nitrocellulose (CAST® and FAST® Slides from Whatman/Schleicher Schnell), in microtitre plate wells or on beads. We have tested a number of surfaces to ensure maximum compatibility with crude rabbit antisera, the printing robot and the assay (in

Table 13.1 Protein markers for use in forensic and biometric applications

(a) Time of Death	
Seasonal	
Winter	IL-10 ↓, IFNa ↑, IFNg ↑
Autumn/winter	IL-6 ↑
Seasonal affective disorder	IL-10 ↓, (IL-6, IFNa, IFNg) ↑, sIL-2R ↑
Diurnal	
Sleep–wake behaviour	Parallels IL-1, TNF
Sleep–wake behaviour	MT ↓, 6-SMT ↓ (↑↑ at night)
Time since death	
Postmortem interval (PMI) marker	Cleavage of C3
PMI up to 24-h marker	Hypoxia-inducible levels of VEGF
PMI (0–5 days) marker	Degradation of cTnI ~ log(time)
Survival time	
<5 min survival time (incised wounds)	IL-1b ↑↑, IL-6 ↑↑, TNFa ↑↑
Neck injury: few min survival time	MG ↑↑
Neck injury: several min survival time	FN ↑
Neck injury: long time	(C5b, C6, C7, C8) ↑

(b) Lifestyle	
Diet	
Carbohydrate, polyunsaturated fat, animal protein, milk, dairy products, fish, poultry, minerals (K, Zn, Mg, Ca, P)	IGF-I ↑
High vegetable consumption, particularly tomatoes	IGF-I ↓
Western diet (red and processed meats, sweets, desserts, French fries, refined grains)	CRP ↑, IL-6 ↑, ES ↑, ICAM-1 ↑, VCAM-1 ↑ insulin ↑, C-peptide ↑, Lp ↑, HC ↑, F ↓
Prudent diet (higher intake of fruit, vegetables, legumes, fish, poultry, whole grains)	CRP ↓, ES ↓, F ↑, insulin ↓, HC ↓
Mediterranean diet	CRP ↓, IL-6 ↓, HC ↓, Fb ↓
Coffee consumption	IL-6 ↑, CRP ↑, SAA ↑, TNFa ↑
Moderate and severe malnutrition	IL-6 ↑, Alb ↓, Tr ↓
Protein malnutrition	STP ↓
Diet, weight loss	CRP ↓
Alcohol consumption	
Alcohol intake (in healthy individuals and alcoholic liver disease)	(IL-6, IL-8) ↑, (IL-4, IL-10) ↑, TGFb ↑, PGE2 ↑, IL-12 ↓, IFNg ↓, IL-1_?, TNF_?
Chronic alcohol abuse	CDT ↑
Alcoholics	Lp ↓
Alcohol abstinence	IL6 ↓, IL10 ↓, IL8 ↓
Smoking	
Exposure to either cadmium or tobacco smoke	Hsp70 ↑
Smoking, intensive smoking	CRP ↑, IL-13 ↑, SHBG ↑
Smoking cessation	CRP ↓

Table 13.1 *Continued*

(b) Lifestyle	

Sex life	
Coitus-induced orgasm (both males and females)	Prl ↑ (remains ↑ for 1 h)
Higher probability of having baby daughters	If HLA-B15 allele present
Behaviour status	
'Sickness behaviour' (fever, weakness, malaise, listlessness, inability to concentrate, depression, lethargia, anhedonia and loss of appetite)	(IL-1, IL-6, IFNs, TNFα) ↑
Panic patients (social phobia)	HLA-DR ↑ (HLA-DR ↓)
Prolonged wakefulness	IL-1 ↑, TNF ↑
Behavioural changes (anorexia, adipsia, sleepiness and depression in social, sexual and general activity)	Pro-inflammatory cytokines ↑
Sexual arousal, aggression, emotional tone, cognition, dominant behaviour (especially violent offenders)	Testosterone ↑, (↑↑); see below for association with HLA alleles
High testosterone (men/women)	B12, 14, 18, 27, DR4, DR1/B12, B5
Low testosterone (men/women)	B15/B8, DR4
Depression	(IL-1b, IL-6, TNFα) ↑
Major depression	CRH ↑
State of health	
Higher risk of cardiovascular disease and diabetes	IGF-I ↓
Higher risk of cancer	IGF-I ↑
Recent exposure to environmental factors: hypothermia, heavy metals, amino acid analogues, inhibitors of energy metabolism, UV exposure; pathological conditions: viral infection, fever, inflammation, ischaemia, oxidative stress	Hsps ↑, cytokines ↑, growth factors ↑
Ongoing inflammatory process prior to death	CRP ↑
Socioeconomic status	
For under 60 years old, 'status' correlates with	CRP ↑, Fb ↑, SAA (↑)
Low educated individuals	CRP ↑
Physical activity	
Leisure time physical activity	CRP ↓
High physical activity (>7 kcal/min expended)	CRP ↓, SAA ↓, TNFα ↓, Fb ↓, IL6 ?
Acute training response	IL-6 ↑
Extreme physical exercise	IL-1 ↑, IL-6 ↑↑, IL-8 ↑, IL-10 ↑↑
Muscle damage and repair	TNFα ↑, stress proteins (e.g. Hsp70) ↑

Table 13.1 *Continued*

(c) Biometrics	
Anthropometric factors	
Adult height	sIGF-I ↑
Leg length, nutrition	Fb ↓
Trunk length	Fb (↓)
Waist/hip ratio	IGF-I ↑; IGFBP-3 ↑; Lp ↓
Hair, baldness	
Hair loss	(IFNg, IL-2, TNFa) ↑
Alopecia areata	MHC HLA class II (*DR4, DR11, DQ*03*) ↑
	HLA III (Notch 4) ↑↑; type 1 cytokines ↑
Age	
Age correlation	HG ↑; Lp ↑; CRP ↑; IGF-I (×30 over life) ↓
Age 50+	MG ↑
Body mass	
Upper body obesity, adiposity	IGF-I ↑; IGFBP-3 ↑
Obese individuals	(TNF, sTNF-RII, IL-6, CRP) ↑; Adn ↑
Body mass index	HG ↑; Lp ↑; CRP ↑; Lp ↑
Sex- and race-related differences	
Men vs. women	Lp (ng/ml): (16.5 +/− 0.3 and 5.7 +/− 0.1)
Black vs. white (women)	Lp (ng/ml): (20.2 +/− 0.6 and 13.9 +/− 0.4)
Black population	cTnI ↑; cTnT ↑
Men vs. women	MG: high/low

(d) Trauma and Death	
Antemortem stress, depression, suicide	
Unexpected deaths	Prl ~533 mU/l (at necropsy)
Postoperative deaths or chronic disease	Prl ~1027 mU/l (at necropsy)
Suicide	Prl ~1398 mU/l (at necropsy)
Suicidal depression	IFN ↑, IL-4 ↓, IL-5 ↓, MHC ↑
Non-suicidal depression	IFN ↓, IL-4 ↑, IL-5 ↑
Time since trauma/wound age	
<30 min	FN ↑, cathepsin D ↑, D-Dimer ↑, TNFa ↑, IL-1b ↑ leukotriene B4 ↑, P-selectin ↑ (peak expr. all)
>1 hour	ES ↑ (peak expression)
>1.5 hours	ICAM-1 (CD54) ↑ (peak expression)
>2 hours	TNFb ↑ (peak expression)
>3 hours	VCAM-1 (DD106) ↑ (peak expression)
>4 hours	IL-1a ↑ (peak expression)
>24 hours	IL-6 ↑ (peak expression)
200 hours	IL-2 ↑ (peak expression)
2 weeks	HSP70 ↑
4 weeks	IL-11 (slow ↓ after initial increase)
Survival time in fatal injuries	CRP ↓
Burn injury 2–9 days	IL-10 ↑ (×3, days 2–9)

Table 13.1 *Continued*

(d) Trauma and Death	
Burn injury 9–14 days	IL-10 ↓ (×1.5, days 9–14 following initial increase)
Severity of trauma	
Damage received before death	IL-8 ↑
Fatality from burns (probability)	IL-8 ↑
Severity of injury and mortality rate	Pro-inflammatory cytokines ↑
Degree of tissue injury (fatal cases)	CRP ↑
Severity of the injury	C3C ↑, C3aC ↑
Prognosis/mortality	(C3, C3a) ↓
Skeletal muscle damage, exertional muscle hyperactivity, convulsive disorders + hypoxia	Postmortem urinary MG ↑
Severity of chest trauma	IL-6 ↑
Extent of overall soft-tissue trauma	IL-6 ↑, IL-8 ↑
Polytraumatized patients: survivors	IL-11 ↑, then decreasing within 4 weeks
Polytraumatized patients: non-survivors	IL-11 ↑, continuously growing
Burns	
Thermal injury	IL-10 ↑
Severe burn injury	IL-10 ↑, type-I APPs (a1AG, CRP, C3C) ↑ Type-II APPs (a2MG, a1AT, HG, Fb) ↑
Severe burn injury	HLA-DR ↓, Alb ↓, pre-Alb ↓, Tr ↓, RBP ↓
Systemic inflammatory response syndrome	TNFa ↑, IL-1 ↑, IL-6 ↑, IL-8 ↑, TGFb ↑
Fast response to burn injury	Prl ↑
Sepsis	
Severe sepsis	HLA-DR ↓
Injury plus infection	HLA-DQ ↓
Septic death in burned patients	IL-6 ↑, TGFb ↓, IL-10 (no change)
Postmortem diagnosis of sepsis	sIL-2R (above 1000 U/ml)
Brain trauma	
Hyperthermia, ischaemia, seizures, traumatic brain injury	Hsp70 ↑ (up to 2 weeks), TNFa ↓
Brain trauma	IL-6 ↑, IL-8 ↑, IL-10 ↑
Miscellaneous trauma	
Abdominal trauma	IL-11 ↑
Stab and incised wounds	IL-1b ↑, IL-6 ↑, TNFa ↑
Drowning not acute myocardial infarction/ischaemia	SP-A ↑
AMI/ischaemia not drowning	cTn-T ↑
Acute respiratory distress	C3a ↑
Polytraumatized individuals	IL-11 ↑, HB-EGF ↑
(e) Blood origin	
Traumatic vs. non-traumatic blood	
Traumatic blood: non-traumatic blood, nasal haemorrhage, menstrual blood	BE = 30 : 1

Table 13.1 *Continued*

(e) Blood origin	
Neonatal vs. adult blood	
Adults	HLA-DR (normal level)
Neonates	HLA-DR ↓
Menstrual vs. peripheral blood	
Menstrual: peripheral bloodstains	FDP, D-Dimer = 200:1
Postmortem: menstrual bloodstains	MG = 4000:1
Menstrual blood marker	MMP ↑

Abbreviations: 6-SMT, 6-sulphatoxymelatonin; a1AG, α_1-acid glycoprotein; a1AT, α_1-antitrypsin; a1FP, α_1-fetoprotein; a2HSG, α_2-HS glycoprotein; a2MG, α_2-macroglobulin; A-diol-g, 17β-androstanediol glucuronide; Adn, adiponectin; Alb, albumin; ApoA1, apolipoprotein A-1; ApoA2, apolipoprotein A-2; ApoB100, apolipoprotein B-100; ApoB48, apolipoprotein B-48; APPs, acute phase proteins; BE, beta-enolase; C1i, C1 inhibitor; C3aC, protein complement C3a; C3C, protein complement C3; C4bbp, C4b binding protein; (C5b-9) assembly of complement plasma glycoproteins; CDT, carbohydrate-deficient transferrin; CKMB, creatine kinase-MB; CN, calcineurin; CPn, ceruloplasmin; CRH, corticotrophin-releasing hormone; CRP, C-reactive protein; cTn, cardiac troponins; D, defensin; D-Dimer, a breakdown product of fibrin; EGF, epidermal growth factor; ES, E-selectin; F, folate; FacB, factor B; FacH, factor H; Fb, fibrinogen; FDP, fibrinogen degradation products; FH, fetal haemoglobin; FLRG, follistatin-related gene; FN, fibronectin; GASP1, growth and differentiation factor-associated protein-1; GDF8mp, myostatin (mature peptide); GDF8pp, myostatin propeptide; HB-EGF, heparin-binding EGF-like growth factor; HC, homocysteine; HG, haptoglobin; HLA, human leucocyte antigens; HSP, heat shock protein(s); Hx, haemopexin; ICAM, intercellular adhesion molecule(s); IFN, interferon(s); In, insulin; IGFBP, insulin-like growth factor-binding protein; IGF-I, insulin-like growth factor I; IL, interleukin(s); IL-1Ra, interleukin-1 receptor antagonist; Lp, leptin; LSA, liver-specific antigen; MBP, mannan-binding protein; MG, myoglobin; MMP, matrix metalloproteinase(s); MRP, migration inhibitory factor-related protein(s); MT, melatonin; PCT, procalcitonin; PGE2, prostaglandin E2; Pre-Alb, pre-albumin; Prl, prolactin; PSA, prostate-specific antigen; RBP, retinal-binding protein; SAA, serum amyloid-A; SHBG, sex hormone-binding globulin; SI, sucrase–isomaltase; sIL-2R, soluble interleukin-2 receptor; sIL-6R, soluble interleukin-6 receptor; SP-A, pulmonary surfactant-associated protein A; sTNF-RII, soluble TNF receptor II; TGFb, transforming growth factor-beta; TNFa/b, tumour necrosis factor alpha/beta; Tr, transferrin; VCAM, vascular cell adhesion molecule; VEGF, vascular endothelial growth factor.

which the binding of labelled peptides, typically at saturated concentrations, competed with unlabelled samples). Figure 13.2 summarizes some of the most recent findings (see Soloviev and Terret, 2005, for a more detailed description). Based on these results, positively charged nylon appears to be the best substrate for most of the experiments, with the main advantages being high binding capacity and the absence of fluorescence quenching. The disadvantage of having a three-dimensional solid support is the increased washing time, but taking into account that the analyte would be small peptides, this would not have any adverse effect on the assay (as confirmed by our data, not shown).

An example of the affinity peptidomics analysis of blood samples is as follows. In order to simulate a forensic scenario, whole blood samples provided by vol-

Table 13.2 Twenty top forensic and biometric protein markers and their anti-peptide antibodies

Name[a]	Biometrics	Lifestyle	Time of death	Trauma and death	Blood origin	No of conditions	ELISA[b]
MG	y		y	y	y	4	25 000
TNFa	y	y	y	y		4	50 000
CRP	y	y		y		3	100 000
Fb (D_Dimer)	y	y		y		3	100 000
IFNg	y	y	y			3	300 000
IL-1a		y	y	y		3	100 000
IL-1b		y	y	y		3	300 000
IL-2	y			y		2	250 000
IL-8		y		y		2	500 000
IL-10		y	y	y		3	500 000
BE				y	y	2	500 000
ES		y		y		2	50 000
FN			y	y		2	500 000
HG	y			y		2	25 000
HSP 70 family		y		y		2	20 000
IGF-I	y	y				2	100 000
Lp	y	y				2	50 000
Prl !!!		y		y		2	20 000
TGFb1		y		y		2	200 000
TGFb4		y		y		2	500 000

[a] For abbreviated protein names, see the legend to Table 13.1.
[b] ELISA results are an indicator of the affinity and the titre of the antibody in sera samples obtained in our studies. The values are determined by diluting the sera samples until ELISA staining reaches the background level.

unteers were transferred onto filter paper discs, dried and stored under different environmental conditions. Following a 9-month storage period under the conditions outlined in Table 13.3, paper discs were incubated with trypsin/Tween-20 for 24 hours at 37°C to achieve solubilization and digestion of the samples. Inactivation of trypsin by heating was followed by the addition of protease inhibitors. The samples were used in the competitive displacement assay in which either labelled synthetic peptides or labelled pooled human tryptically digested plasma were used. It was possible to match some of the five samples in most cases tested, except for the samples from two batches, most probably due to complete loss of the proteinaceous matter. The sensitivity of detection can be improved by reducing the reaction volume (in the above experiment, 5 µl of whole blood was used per 1-ml assay). Table 13.4 presents some of the examples in which all five different simulated forensic samples were matched correctly (with or without false positives). Pearson's correlation coefficient was calculated for each pair of arrays (identically made arrays, assayed with differently stored samples) and was used as an indicator of matching (or mismatching)

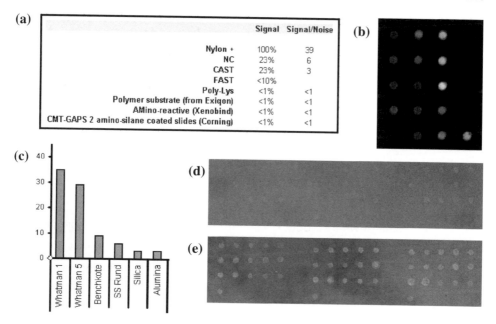

	Signal	Signal/Noise
Nylon +	100%	39
NC	23%	6
CAST	23%	3
FAST	<10%	
Poly-Lys	<1%	<1
Polymer substrate (from Exiqon)	<1%	<1
AMino-reactive (Xenobind)	<1%	<1
CMT-GAPS 2 amino-silane coated slides (Corning)	<1%	<1

Figure 13.2 Evaluation of anti-peptide affinity array production parameters. (a) Choice of substrate: data obtained by spotting fluorescently labelled albumin onto untreated substrates. The values reflect not only the protein binding capacity of the substrate but also the fluorescence quenching by the substrate. (b) An example of the best substrate: Nylon membrane, immobilized onto a glass slide. (c) Fluorescence quenching by other (non-traditional) substrates. (d) The effect of crosslinking on signal stability (following blocking and washing. (e) Untreated membrane: membranes were incubated in sealed beakers filled with undiluted formaldehyde (10 ml per 100-ml beaker volume) overnight

samples. Note that fewer antibodies could have been used to achieve a correct match.

The effect of sample collection and storage conditions on the sample stability has been discussed in the literature for many years, mostly in relation to clinical chemistry applications. Affinity peptidomics is the most generic affinity assay system to date, applicable for protein identification, quantification and expression profiling. In its concept it is the opposite of the 'sample preservation' strategies employed or attempted in traditional biomolecular diagnostics. It requires complete proteolytic digestion of the proteinaceous samples as the first step of any analysis. It is therefore suitable for the analysis of inconsistently stored or partially degraded samples, hence is suitable for a wide range of application, ranging from routine biomedical diagnostics to biometrics and forensics analyses. Protein microarrays and peptidomics are the two enabling technologies that allow fast and accurate protein profiling and would yield a new alternative methodology that is likely to supersede traditional protein

Table 13.3 Simulation of forensic material (samples were stored for 9 months under the conditions specified)

Batch no.[a]	Storage	To simulate
001–003	×50 µl of whole blood, dried onto paper filters, stored at +4°C	A drop of blood
006	×50 µl of whole blood, dried onto paper filters, stored at room temperature	A blood spot on clothing (stored)
007	×50 µl of whole blood, dried onto paper filters, stored at +37°C	A blood spot on clothing (worn)
008	×50 µl of whole blood, dried onto paper filters, stored at the temperature cycling conditions (daily cycles of between 15°C and up to approx 50°C)	A blood spot on clothing (exposed to heating/cooling)
011	×50 µl of whole blood, dried onto paper filters, stored at room temperature, ironed prior to use	A blood spot on clothing (ironing)
013	×50 µl of whole blood, dried onto paper filters, stored at room temperature, incubated in detergent prior to use	A blood spot on clothing (washing)
015	×50 µl of whole blood, dried onto paper filters, stored at room temperature, on a windowsill, facing south (full sun)	A blood spot on clothing (exposed to sun)

[a] Each batch contained five different (matching) samples, dried onto individual paper filters.

Table 13.4 Sample matching using affinity peptidomics assays

(a) Samples stored at 4°C vs. room temperature[a]

	Sample 1	Sample 2	Sample 3	Sample 4	Sample 5
Sample 1	**0.89**	−0.79	−0.09	0.71	−0.31
Sample 2	−0.88	**0.94**	0.08	−0.81	−0.14
Sample 3	−0.04	−0.71	**0.83**	0.81	0.68
Sample 4	0.37	−0.88	0.63	**0.97**	0.19
Sample 5	0.07	−0.56	0.26	0.50	**0.95**

(b) Samples stored at 4°C vs. samples having undergone temperature cycles

	Sample 1	Sample 2	Sample 3	Sample 4	Sample 5
Sample 1	**0.96**	−0.23	−0.56	−0.88	−0.34
Sample 2	−0.07	**0.55**	−0.55	−0.22	0.90[b]
Sample 3	−0.33	−0.35	**0.75**	0.60	−0.78
Sample 4	−0.74	−0.12	0.83	**0.90**	−0.39
Sample 5	0.11	−0.25	−0.28	−0.33	**0.80**

[a] All five samples matched correctly. As little as four antipeptide antibodies were sufficient (IL-1, ES, FN, SP-A) to match all five samples.
[b] This is a single false-positive match (FN, Prl, TGF).

analysis techniques. There appears to exist a sufficient number of protein markers highly relevant to the forensics and biometrics fields. We have listed ~ 100 such markers (see above) and for many of these some quantitative data already exist regarding their up/down-regulation (lifestyle, trauma, disease, etc.). A serious meta-analysis of the literature is necessary for transforming the outcome of each study into a 'common currency', a measure of the effect size, which reflects the magnitude of the effect and could be compared across studies. Otherwise, existing literature data cannot be used reliably. However, even with a conservative estimate of a twofold expression difference for each marker (there are examples of ×4000-fold differences) and 25% error in quantification, 100 different protein markers may yield very approximately up to 4^{100} different combinations (or expression states) of these markers. This simple estimate shows the huge potential of protein-based diagnostics and especially its application to forensics and biometrics.

13.5 References

Ahlstedt, S., Kristonffersson, A., Pettersson, E. and Svard, P.O. (1976) Binding properties of rabbit antibodies against various penicillins as studied by immunoprecipitation, indirect hemagglutination and radiometric immunoassay. *Int. Arch. Allergy Appl. Immunol.* 51: 145–155.

Attili, B.S. and Suleiman, A.A. (1996) A piezoelectric immunosensor for the detection of cocaine. *Microchem. J.* 54: 174–179.

Ben Ali, M., Korpan, Y., Gonchar, M., El'skaya, A., Maaref, M.A., Jaffrezic-Renault, N. and Martelet, C. (2006) Formaldehyde assay by capacitance versus voltage and impedance measurements using bi-layer bio-recognition membrane. *Biosens. Bioelectron.* 22: 575–581.

Barry, R., Diggle, T., Terrett, J. and Soloviev, M. (2003) Competitive assay formats for high-throughput affinity arrays. *J. Biomol. Screen.* 8: 257–263.

Barry, R. and Soloviev, M., (2004) Quantitative protein profiling using antibody arrays. *Proteomics* 4: 3717–3726.

Cagney, G. and Emili, A. (2002) De novo peptide sequencing and quantitative profiling of complex protein mixtures using mass-coded abundance tagging. *Nat. Biotechnol.* 20: 163–170.

Clark Jr., L.C. and Lyons, C. (1962) Electrode systems for continuous monitoring in cardiovascular surgery. *Ann. NY Acad. Sci.* 102: 29–45.

Cooper, M.A., Dultsev, F.N., Minson, T., Ostanin, V.P., Abell, C. and Klenerman, D. (2001) Direct and sensitive detection of a human virus by rupture event scanning. *Nat. Biotechnol.* 19: 833–837.

D'Orazio, P. (2003) Biosensors in clinical chemistry. *Clin. Chim. Acta* 334: 41–69.

Delmulle, B.S., De Saeger, S.M.D.G., Sibanda, L., Barna-Vetro, I. and Van Peteghem, C.H. (2005) Development of an immunoassay-based lateral flow dipstick for the rapid detection of aflatoxin B-1 in pig feed. *J. Agric. Food Chem.* 53: 3364–3368.

Dickson, E.F.G., Pollak, A. and Diamandis, E.P. (1995) Ultrasensitive bioanalytical assays using time-resolved fluorescence detection. *Pharmacol. Ther.* 66: 207–235.

Drafts, B. (2001) Acoustic wave technology sensors, retrieved 13 July 2006 from *http://www.sensorsmag.com/articles/1000/68/main.shtml.*

Du, J., Harding, G.L., Ogilvy, J.A., Dencher, P.R. and Lake, M. (1996) A study of Love-wave acoustic sensors. *Sens. Actuat. A* **56**: 211–219.

Dzyadevych, S.V., Soldatkin, A.P., El'skaya, A.V., Martelet, C. and Jaffrezic-Renault, N. (2006) Enzyme biosensors based on ion-selective field-effect transistors. *Anal. Chim. Acta* **568**: 248–258.

Garden, S.R. and Strachan, N.J.C. (2001) Novel colorimetric immunoassay for the detection of aflatoxin B-1. *Anal. Chim. Acta* **444**: 187–191.

Gerber, S.A., Rush, J., Stemman, O., Kirschner, M.W. and Gygi, S.P. (2003) Absolute quantification of proteins and phosphoproteins from cell lysates by tandem MS. *Proc. Natl. Acad. Sci. USA* **100**: 6940–6945.

Gizeli, E., Goddard, N.J., Lowe, C.R. and Stevenson, A.C. (1992) A Love plate biosensor utilizing a polymer layer. *Sens. Actuat. B* **6**: 131–137.

Gizeli, E., Liley, M., Lowe, C.R. and Vogel, H. (1997) Antibody binding to a functionalized supported lipid layer: A direct acoustic immunosensor. *Anal. Chem.* **69**: 4808–4813.

Gizeli, E., Stevenson, A.C., Goddard, N.J. and Lowe, C.R. (1993) Acoustic Love plate sensors – comparison with other acoustic devices utilizing surface SH-waves. *Sens. Actuat. B* **14**: 638–639.

Gygi, S.P., Rist, B., Gerber, S.A., Turecek, F., Gelb, M.H. and Aebersold R. (1999) Quantitative analysis of complex protein mixtures using isotope-coded affinity tags. *Nat. Biotechnol.* **17**: 994–999.

Haupt, K. and Mosbach, K. (1998) Plastic antibodies: developments and applications. *Trends Biotechnol.* **16**: 468–475.

Khosravi, M.J., Papanastasioudiamandi, A. and Mistry, J. (1995) An ultrasensitive immunoassay for prostate-specific antigen based on conventional colorimetric detection. *Clin. Biochem.* **28**: 407–414.

Kovacs, G. and Venema, A. (1992) Theoretical comparison of sensitivities of acoustic shear-wave modes for (bio)chemical sensing in liquids. *Appl. Phys. Lett.* **61**: 639–641.

Kretschmann, E. (1971) Determination of optical constants of metals by excitation of surface plasmons. *Z. Phys.* **241**: 313.

Kronick, M.N. (1986) The use of phycobiliproteins as fluorescent labels in immunoassay. *J. Immunol. Methods* **92**: 1–13.

Kusnezow, W., Jacob, A., Walijew, A., Diehl, F. and Hoheisel, J.D. (2003) Antibody microarrays: An evaluation of production parameters. *Proteomics* **3**: 254–264.

Leonard, P., Hearty, S., Brennan, J., Dunne, L., Quinn, J., Chakraborty, T. and O'Kennedy, R. (2003) Advances in biosensors for detection of pathogens in food and water. *Enzyme Microb. Technol.* **32**: 3–13.

Merwe, P.A., van der (2003) Surface plasmon resonance, retrieved 13 July 2006 from *http://users.path.ox.ac.uk/~vdmerwe/Internal/spr.PDF.*

Moorthy, J., Mensing, G.A., Kim, D., Mohanty, S., Eddington, D.T., Tepp, W.H., Johnson, E.A. and Beebe, D.J. (2004) Microfluidic tectonics platform: A colorimetric, disposable botulinum toxin enzyme-linked immunosorbent assay system. *Electrophoresis* **25**: 1705–1713.

Muratsugu, M., Ohta, F., Miya, Y., Hosokawa, T., Kurosawa, S., Kamo, N. and Ikeda, H. (1993) Quartz-crystal microbalance for the detection of microgram quantities of human serum-albumin – relationship between the frequency change and the mass of protein adsorbed. *Anal. Chem.* **65**: 2933–2937.

Niedbala, R.S., Feindt, H., Kardos, K., Vail, T., Burton, J., Bielska, B., Li, S., Milunic, D., Bourdelle, P. and Vallejo, R. (2001) Detection of analytes by immunoassay using up-converting phosphor technology. *Anal. Biochem.* **293**: 22–30.

Nord, K., Gunneriusson, E., Ringdahl, J., Stahl, S., Uhlen, M. and Nygren, P.A. (1997) Binding proteins selected from combinatorial libraries of an alpha-helical bacterial receptor domain. *Nat. Biotechnol.* **15**: 772–777.

Qian, S.Z. and Bau, H.H. (2003) A mathematical model of lateral flow bioreactions applied to sandwich assays. *Anal. Biochem.* **322**: 89–98.

Qian, S.Z. and Bau, H.H. (2004) Analysis of lateral flow biodetectors: competitive format. *Anal. Biochem.* **326**: 211–224.

Raja, A., Machicao, A.R., Morrissey, A.B., Jacobs, M.R. and Daniel, T.M. (1988) Specific detection of mycobacterium-tuberculosis in radiometric cultures by using an immunoassay for antigen-5. *J. Infect. Dis.* **158**: 468–470.

Rongen, H.A.H., Hoetelmans, R.M.W., Bult, A. and Vanbennekom, W.P. (1994) Chemiluminescence and immunoassays. *J. Pharm. Biomed. Anal.* **12**: 433–462.

Sakai, G., Saiki, T., Uda, T., Miura, N. and Yamazoe, N. (1995) Selective and repeatable detection of human serum-albumin by using piezoelectric immunosensor. *Sens. Actuat.* **B 24**: 134–137.

Sharma, S.K., Eblen, B.S., Bull, R.L., Burr, D.H. and Whiting, R.C. (2005) Evaluation of lateral-flow clostridium botulinum neurotoxin detection kits for food analysis. *Appl. Environ. Microbiol.* **71**: 3935–3941.

Soloviev, M., Barry, R., Scrivener, E. and Terrett, J. (2003) Combinatorial peptidomics: a generic approach for protein expression profiling. *J. Nanobiotechnol.* **1**: 4.

Soloviev, M., Barry, R. and Terrett, J. (2004) Chip based proteomics technology. In: *Molecular Analysis and Genome Discovery* (R. Rapley and S. Harbron, eds), John Wiley & Sons, Chichester, pp. 217–249.

Soloviev, M. and Finch, P. (2005) Peptidomics, current status. *J. Chromatogr., B: Anal. Technol. Biomed. Life Sci.* **815**: 11–24.

Soloviev, M. and Terrett, J. (2005) Practical guide to protein microarrays: assay systems, methods and algorithms. In: *Protein Microarrays* (M. Schena, ed.), Jones and Bartlett, Sudbury, MA, pp. 43–55.

Su, X.D., Robelek, R., Wu, Y.J., Wang, G.Y. and Knoll, W. (2004) Detection of point mutation and insertion mutations in DNA using a quartz crystal microbalance and MutS, a mismatch binding protein. *Anal. Chem.* **76**: 489–494.

Thompson, M., Arthur, C.L. and Dhaliwal, G.K. (1986) Liquid-phase piezoelectric and acoustic transmission studies of interfacial immunochemistry. *Anal. Chem.* **58**: 1206–1209.

Wang, J. (2001) Glucose biosensors: 40 years of advances and challenges. *Electroanalysis* **13**: 983–988.

Yalow, R.S. (1980) Radioimmunoassay. *Annu. Rev. Biophys. Bioeng.* **9**: 327–345.

Yang, C.Y., Brooks, E., Li, Y., Denny, P., Ho, C.M., Qi, F.X., Shi, W.Y., Wolinsky, L., Wu, B., Wong, D.T.W. and Montemagno, C.D. (2005) Detection of picomolar levels of interleukin-8 in human saliva by SPR. *Lab. Chip* **5**: 1017–1023.

14

Application of MRS in forensic pathology

Eva Scheurer, Michael Ith and Chris Boesch

14.1 Forensic, criminalistic and ethical significance of time of death

Knowledge of the postmortem interval (PMI) is absolutely essential in criminal investigations for further measures of tracing, collection of evidence and inclusion or exclusion of suspects. However, the current methods for estimating PMIs – particularly longer ones – are far from accurate and straightforward. Therefore, the boards of prosecution deem a new method that may solve these problems extremely important. Additionally, the procedures and results of forensic examinations have a strong impact on relatives and friends of a victim, influencing the process of mourning and the management of sentiments of real and imagined guiltiness. Questions such as 'would an earlier visit have saved my beloved person?' are often as fundamental as the cause of death itself. The value of a reliable evaluation of the PMI for relatives has been shown in a study (Plattner *et al.*, 2002) where 43% of the interviewed persons wanted to know the circumstances of the unexpected death, including the exact time. It is obvious that a determination of PMI using non-invasive techniques would further reduce the trauma for relatives and friends, resulting in an increased willingness to agree with the forensic investigation. We will report here the estimation of PMIs using magnetic resonance spectroscopy (MRS), a non-invasive method that acquires metabolic data from tissue *in situ*. Since magnetic resonance imaging (MRI) is increasingly used to examine bodies in forensic investigations (Thali *et al.*, 2003; Yen *et al.*, 2004), MRS can be combined with these imaging methods during the same examination.

14.2   Classical methods for the determination of PMI

Traditionally, time of death was defined by an irreversible termination of circulation caused by a cessation of cardiac function. This definition had to be modified in intensive care medicine where circulation and respiration were maintained artificially. The resulting working definition uses brain death as the moment that characterizes individual death (Kurthen *et al.*, 1989 ; Capron, 2001). Nevertheless, cessation of cardiac function still represents the time of death in the vast majority of forensic cases and thus the transition of the body into the so-called supravital state (Madea and Henssge, 1991). Supravital reactions are, by definition, phenomena that can be observed in the time between irreversible loss of function of the organ systems and the actual death of the cells in a particular organ. Since the resistance of various tissues to oxygen deficiency differs greatly, these phenomena in different tissues can cover a time span from a few minutes to some hours following the irreversible loss of brain function.

In the early PMI (Figure 14.1), i.e. during the first 3 days postmortem, estimation of the PMI in forensic medicine is mostly based on the evaluation of supravital signs (e.g. livor mortis, rigor mortis) and the decrease of body temperature after death (Mallach and Mittmeyer, 1971; Krompecher and Fryc, 1979). The medico-legal significance of livor and rigor mortis is limited by nonstandardized and subjective examination techniques (Schleyer, 1975) and various ante- and postmortem factors (Forster *et al.*, 1974; Krompecher *et al.*, 1983; Fechner *et al.*, 1984). The decrease of body temperature after death – alone or in combination with non-temperature-based indicators (Henssge *et al.*, 2000b) –

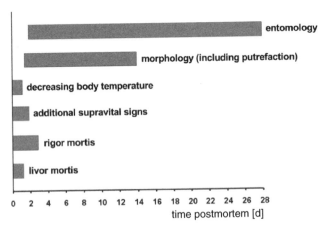

Figure 14.1 Overview of the periods in which the classical forensic methods can be used for estimation of the postmortem interval (PMI). It illustrates that reliable parameters such as decreasing body temperature and rigor and livor mortis are available no longer than the first few days. After that short period, rather subjective and non-standardized methods need to be consulted

is a reliable and thoroughly investigated phenomenon (Marshall and Hoare, 1962; Shapiro, 1965; Marty, 1995; Henssge et al., 2000a). Considering the weight of the body and environmental factors, the body core temperature permits a retro-calculation of a time span within which death occurred. However, body temperature approaches ambient temperature approximately 30 hours after death and thereafter is no longer of value for assessing time of death. A survey at the Institute of Forensic Medicine at the University of Bern revealed that 16% of 559 bodies were not found before two or more days after death. This corresponds to a significant number of cases where an objective method for determination of the PMI is missing. In that later postmortem phase, an estimation of the PMI has to rely on the evaluation of putrefaction signs of the body (Schneider and Riese, 1980), fragmentary criminological information from witnesses or, in certain cases, entomological studies of insects colonizing the body (Figure 14.1).

Putrefaction signs can be observed starting at 2–3 days postmortem, however, while the general sequence of their appearance has been described frequently (Berg, 1975), a review of the forensic literature showed that no systematic description of morphological changes during decomposition of the body giving precise time frames for each sign had been established. This finding was additionally documented by an internal survey including nine experienced forensic pathologists. It was found that even common changes being observed in the majority of cases led to an uncertainty of up to 5 days for PMIs of the same time range. Thus, it is generally accepted that morphological signs cannot be used as reliable and objective indicators of the PMI (Schneider and Riese, 1980), particularly since they are strongly dependent on external (e.g., temperature, humidity) and internal factors (e.g., antemortem treatment and diseases) (Mann et al., 1990). Forensic entomology uses the fact that insects colonize bodies during the process of decomposition in various waves (Smith, 1986), depending on the stage of decomposition of the corpse. While this method allows for an estimation of the PMI in certain cases (Benecke, 1996), it is extremely time-consuming and demanding. In addition, insect populations vary with geographical region, season and environment (Campobasso et al., 2001; Grassberger and Reiter, 2001), making standardization almost impossible.

The disintegration of the chemical, physical and morphological organization starting immediately after death is known as 'autolysis' and results from the action of endogenous enzymes and the cessation of oxygen-dependent biochemical processes. The failure of the body to maintain homeostasis leads to a breakdown of the internal equilibrium as well as to the degradation of proteins, carbohydrates and fats. This results in an increase of breakdown products. Chemical analyses for the purpose of a PMI estimation based on concentration changes have been performed on body fluids such as blood and blood serum (Coe, 1974), cerebrospinal fluid (Endo et al., 1990), vitreous humor of the eyes (Coe, 1989), synovial fluid (Madea et al., 2001) and skeletal muscle (Mayer and Neufeld, 1980; Mittmeyer, 1980). Some projects concentrated on specific groups of metabolites, e.g. the decomposition of fats (Lindlar, 1969; Doring, 1975) or proteins (Bonte *et al.*, 1976; Bonte, 1978). Collection of body fluids can be very difficult depending on the PMI and the state in which the individual body is found (Henssge and Madea, 1988). While this could be less problematic in brain tissue due to the protection by the skull, only a few studies were published (Diessner and Lahl, 1969; Daldrup, 1983, 1984) that analysed brain tissue chemically for an estimation of the PMI. On the other hand, examinations of the brain would be particularly interesting because inter-individual differences in tissue composition are very small compared to other tissues of the body (e.g. skeletal muscle or liver).

To date, none of the chemical methods have become generally accepted, nor are they routinely applied for PMI estimation. One reason for this disappointing outcome may be the fact that most of these methods attempted to characterize PMI by a few or even just one specific parameter. It seems that chemical methods based on multiple metabolite concentrations lead to more successful PMI estimations while methods based on a single metabolite failed to be used regularly in the past.

14.3   Magnetic resonance spectroscopy

It would be far beyond the scope of this chapter to explain magnetic resonance in depth. Several textbooks and overviews have been published that give an introduction into the basics of MRS (Gadian, 1982; Boesch, 1999, 2005; De Graaf, 1999).

Magnetic resonance became an established tool in diagnostic radiology and clinical research. Since it is so popular in medicine, it is often ignored that clinical magnetic resonance is just one of many applications of the 'nuclear magnetic resonance' (NMR) effect. In chemistry and biophysics, so-called 'high-resolution' NMR spectroscopy is one of the most important methods to study the structure of organic and inorganic substances. Today, the development and analysis of chemical compounds is unthinkable without NMR technology; even complicated three-dimensional structures of large biological molecules can be understood with the help of this potent method. In medicine, magnetic resonance imaging (MRI) is nowadays a particularly valuable and versatile instrument in diagnostic radiology (Glover and Herfkens, 1998), due to its detailed and accurate representation of *in vivo* anatomy and function. Magnetic resonance spectroscopy (MRS) combines the volume-selective data acquisition of MRI with the chemical information provided by NMR. Without reaching the sensitivity and resolution of high-resolution NMR, *in situ* MRS allows for a non-invasive examination of the chemical composition from selected volumes in humans (Howe *et al.*, 1993; Boesch, 1999; De Graaf, 1999; Ross and Danielsen, 1999).

In principle, the above-mentioned applications of the NMR effect are all based on the fact that stable atomic nuclei have a magnetic moment and an angular momentum, called 'spin'. When a material is placed in an external magnetic field, these 'spins' begin to tumble in a precession. This motion enables the nuclei to absorb and emit electromagnetic waves with a frequency in the MHz range, depending on the strength of the magnetic field and the type of isotope. While MR images are generated in the majority of cases from the signals of hydrogen nuclei (^{1}H) in water, MR signals can also be obtained from various other stable isotopes, including phosphorus (^{31}P), carbon (^{13}C), fluorine (^{19}F), sodium (^{23}Na) and others. Magnetic resonance spectroscopy makes particular use of ^{1}H, ^{13}P and ^{31}C. Depending on the local chemical environment, different atoms in a molecule resonate at slightly different frequencies, resulting in the so-called 'chemical shift'. This chemical shift is resolved in spectroscopic applications (NMR and MRS), however it is neglected in standard imaging. It is measured in relative units (parts per million, ppm), and represents the x-axis of a spectrum. Figure 14.2 illustrates the spectrum of butyrate, a molecule that has three types of hydrogen atoms: an H–C–H group close to the carboxyl group C=O, a second H–C–H group in the middle and a CH$_3$ group at the end of the molecule. The hydrogen atoms in each of these groups are identical and resonate at a specific position on the chemical shift axis. Since the area of the resonance peak under specific experimental conditions is proportional to the concentration of the chemical species, butyrate should show three single resonance lines with a 2:2:3 ratio of the areas. The reason why the lines in Figure 14.2 are further split into so-called 'multiplets' is a mutual influence of the nearest hydrogen atoms. This effect complicates the spectrum; however, it also helps to identify chemical species. In other words, spectroscopy (either NMR or MRS) uses frequency information of absorbed and emitted radio waves to identify different chemical compounds – the position of the signal in the

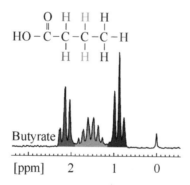

Figure 14.2 An MR spectrum reveals information on the chemical identity (position on the horizontal axis) and the concentration (area under the resonance) of metabolites. The example shows three different chemical groups within butyrate, each group with a specific position on the chemical shift axis. The splitting of the resonance lines is an effect of the number of nearest neighbours in a molecule and can additionally be used to identify a resonance

spectrum defines the chemical nature while the signal area reveals the amount of molecule. The width of a resonance line is another important experimental factor since it defines the resolution of the spectrum, i.e. how many chemical species can be observed separately. In addition to the type of chemical compound, the homogeneity of the magnetic field influences the width of the lines, i.e. the resolution of a spectrum. The homogeneity of the magnetic field can be improved by technical means, called 'shimming', in turn biological tissue can introduce additional inhomogeneity of the field, particularly at air–tissue borders. As we will see below, gas bubbles that are produced during the process of body decomposition can severely decrease the homogeneity of the magnetic field such that spectral resolution becomes a problem.

While high-resolution NMR uses small sample tubes in vertical magnets up to a magnetic field strength of 22 Tesla, *in situ* and *in vivo* MR spectra are acquired in horizontal bore magnets with typical field strengths of 1.5 or 3 Tesla. Selection of the signal-generating tissue ('voxel', 'region of interest ROI') is achieved by techniques that are common with MRI. The voxel can be placed at the desired anatomical location based on a series of localizer MR images (see Figure 14.3 opposite) and contains typically a few millilitres in ^{1}H-MRS. An MR spectrum can be acquired by technicians in much less than an hour without the need for sample preparation and a 1H-MR-spectrum contains information on about 20 metabolites simultaneously. Depending on the automation of the spectral analysis, final data can be obtained in less than an hour.

In order to estimate the PMI, ^{1}H-MRS of the brain is particularly advantageous:

1. ^{1}H-MRS provides high relative sensitivity, allowing the selection of small voxels.

2. Established MRS sequences for ^{1}H allow for a robust and easy selection of spectroscopic volumes with subsequent absolute quantitation of the metabolites.

3. A majority of clinical MRS investigations are performed on the brain, resulting in widespread experimental experience and a large collection of reference data in healthy and pathological brain tissue.

4. About 20 different characterized substances can be observed and quantified.

5. Inter-individual differences in brain tissue composition are rather small (approx. 5%; Kreis, 1997) and can partly be attributed to the aging process (Chang *et al.*, 1996), which can be corrected for.

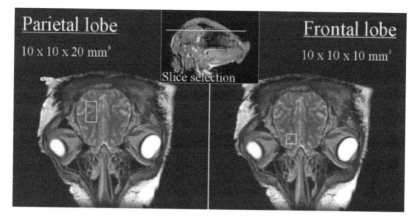

Figure 14.3 Magnetic resonance images of a sheep brain with a placement of voxels for the spectroscopic data acquisition. The images are so called T2-weighted, i.e. the signal acquisition is optimized such that the contrast between different types of tissue is optimal. During decomposition of the brain, voxel placement has to avoid gas bubbles

6. The brain is protected by the skull after death, i.e. destruction by environmental factors (scavengers, external microorganisms) is minimized, and when the brain tissue becomes decomposed and liquefied the skull guarantees a certain fixation.

A determination of late PMIs by means of ^{1}H-MRS of the brain is based on the fact that MRS reveals concentrations of multiple metabolites in a single spectrum. If there are calibration curves of metabolite changes following death, it should be possible to compare the measured concentrations in a brain with an unknown time of death with the calibration curves. The following paragraphs will illustrate how calibration curves are established in an animal model, how first cases of human bodies are studied and which statistical procedures are used to compare actual metabolite concentrations and calibration curves.

Sheep model

Ethical reasons prevent the storage of human bodies longer than necessary for a regular forensic examination, making it almost impossible to establish calibration curves of metabolite changes over a longer time based on human bodies. An animal model allows decomposition processes of the brain to be followed repeatedly and under standardized conditions at different, optimally spaced points in time.

While in the following study a sheep model is used, there have also been reports on studies in pigs (Banaschak *et al.*, 2005). For practical reasons, the animal has to be a mammal and should be available in a slaughterhouse. The size of the animal brain should be as large as possible to optimize voxel selection; however, the complete head should still be within the size of an MR head-coil. As described in detail by Ith *et al.* (2002), the heads used in the following study were harvested from healthy sheep, which died in the course of the normal slaughtering process. The heads were separated from the bodies and stored at constant temperature (21 ± 3°C). *In situ* brain spectra from the frontal lobe and the parieto-occipital region (Figure 14.3) were acquired regularly up to 18 days postmortem. A striking increase of the signal amplitude shows how decomposition of brain tissue makes many more metabolites MR-visible (Figure 14.4). Figure 14.4 also illustrates that other chemical compounds are formed during decomposition.

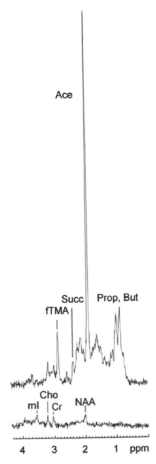

Figure 14.4   Two spectra of sheep brain at the same scale, the lower trace recorded *in vivo* and the upper trace 18 days postmortem. Following death, new metabolites appear and the concentration of visible metabolites increases dramatically in the decomposing brain tissue

In order to obtain quantitative data, absolute concentrations of the metabolites had to be determined. This is achieved by a comparison of the signals with the signal from water (fully relaxed) from the same volume and with an estimation of the resonance area by a mathematical modelling of the experimental curve, i.e. by a fitting procedure. A well-known fitting algorithm is 'LC Model' (Provencher, 1993) that uses linear combinations of model-spectra, a so-called 'basis set' of spectra from measurable substances. The basis set, which in clinical practice is used to fit brain spectra, consists of about 25 metabolites: acetate, alanine, aspartate, cholines, creatines, phospho-ethanolamine, ethanol, gamma-aminobutyric acid, glucose, glutamine, glutamate, glutathione, glycine, lactate, myo-inositol, scyllo-inositol, N-acetyl-asparates, taurine, and valine. However, during the particularly interesting late postmortem phase, several substances appeared that had not been observed in brain spectra before then, and had to be identified by means of high-resolution NMR as trimethylamine, propionate, butyrate and isobutyrate (Ith *et al.*, 2002). Since the additional metabolites are detected only after the third day postmortem and do not contribute significantly to spectra before then, they are most likely of bacterial origin. A chemical identification of substances is not only necessary for a deeper understanding of the underlying processes, but also to analyse the spectra mathematically. An incomplete set of theoretical spectra results in a poorer fit, and spectral features could be wrongly attributed to substances that are included in the basis set. This would lead to an underestimation of the unknown substance and, subsequently, to an erroneous overestimation of an included substance with similar spectral features. Therefore, a complete set of spectra improves the accuracy for all substances in a spectrum. Inclusion of the newly identified metabolites and succinate (corrected for pH and temperature effects) resulted in a significantly improved fit quality and reliable quantitation of up to 30 metabolites.

The time course of the observed metabolites has then been followed up to 18 days postmortem. While some of the time courses are ambiguous or too scattered, others show a clear and unequivocal time dependence, such as NAA/NAAG or butyrate as shown in Figure 14.5. Obviously, the decomposition process follows certain rules that can be used for an estimation of the PMI. In the early postmortem phase metabolic changes probably due to autolytic processes are observed, i.e., a decay of the 'neuronal marker' N-acetyl-aspartate (Figure 14.5a) into acetate and aspartate and the build-up of lactate during glycolysis. Later, starting at about 3 days postmortem, bacterial decomposition processes, i.e. heterolysis, come up in addition to autolysis, leading to an observation of substances that are associated with the action of bacteria, e.g. butyrate (Figure 14.5b), trimethylamine or propionate.

14.4 How to predict the PMI based on MRS measurements

Following the acquisition of MR data, the time course of the metabolite concentrations has to be described by mathematical functions (Figure 14.6, step 1).

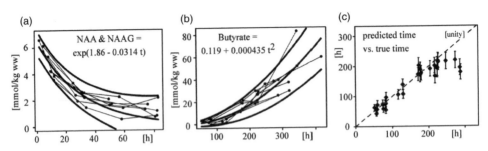

Figure 14.5   (a, b): Calibration curves for two different metabolites together with a comparison of calculated and true times of death. The experimental data are fitted with functions that are as simple as possible. The fitting procedure also gives error ranges of the calibration curve (upper and lower curve): (a) shows the decay of N-acetyl-aspartate (NAA) and N-acetyl-aspartylglutamate (NAAG) due to auto-lytic degradation; (b) demonstrates the increase of butyrate, most likely due to bacterial metabolism. The scattering of the data is surprisingly small (Scheurer et al., 2003). (c) Comparison of predicted times versus true times. In the sheep model, this comparison proves the feasibility of PMI estimations up to 250 hours postmortem. Reproduced with permission from Scheurer et al. (2005)

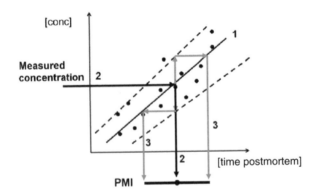

Figure 14.6   Schematic representation of the mathematical calculation of the predicted PMI. The time course of the experimental data points is described by a mathematical function (solid line, step 1). In addition, this fitting procedure of the scattered experimental points leads to error ranges (broken lines). Subsequently, a measured concentration of a metabolite (black arrows, step 2) is projected on the calibration curve, resulting in a predicted PMI. The error range for the PMI is obtained as indicated by the grey arrows (step 3)

Since the exact biochemical mechanisms are not yet known, it is reasonable to chose functions with the smallest number of parameters that still lead to an appropriate description of the time course (Scheurer *et al.*, 2005). Two examples of the concentration changes over time with applied model functions and corresponding confidence intervals are shown in Figures 14.5a and 14.5b. The concentration of NAA+NAAG is described by a decreasing exponential function – it may serve as an indicator for the period below 70 h. Butyrate starts to increase at about 50 h postmortem and reveals an unequivocal function up to 400 h that is parameterized by a quadratic function.

As soon as such a calibration curve is established, a measured metabolite concentration (Figure 14.6, step 2) corresponds to a specific point in time when the metabolite is expected to have this concentration. Figures 14.5a and 14.5b illustrate that a description of the experimental data by mathematical functions allows a definition of error ranges. These error ranges can now be used to estimate the error of the resulting PMI, as shown in the third step of Figure 14.6. It is obvious that equivocal, scattered or flat parts of the time course cannot be used as a calibration curve. Therefore, from the 30 visible metabolites, just about 10 are fitted by mathematical functions and 5 of them are used for further analysis (Scheurer et al., 2005).

Since experimental data from different metabolites will not predict exactly the same PMI, it is necessary to combine the predictions obtained from different metabolite curves by a robust procedure. In addition, and in order to restrict the fit to meaningful (i.e. unequivocal) parts of the measured time course, limits for concentrations and times are applied – full details about this procedure can be found in Scheurer et al. (2005) but to summarize: final estimations for the PMI are calculated on the basis of PMIs from the different metabolites, weighted by their accuracy as determined by Figure 14.6 (step 3).

Figure 14.5c compares, for every measurement time, predictions combined from five metabolites (acetate, alanine, trimethylamine, butyrate and propionate) weighted according to their variances with true PMIs (regression analysis below 250 h: $y = 0.898x + 6.5$). The correlation coefficients of the predicted time versus true PMI are $r = 0.93$ for the whole time period (0–300 h) and $r = 0.97$ for the time period below 250 h. Predicted times combined from these five metabolites correlate very well with true times postmortem up to 250 h. In contrast, PMIs of >250 h are systematically underestimated in this model system. It is surprising that this result is almost independent for various combinations of metabolites ($r = 0.87$–0.97, mean 0.92), showing that the influence of the choice of metabolites is almost negligible. Obviously, a larger number of metabolites leads to smaller variances, however the robustness of the method is convincing. Eventually, a combination of acetate, alanine, butyrate, trimethylamine and propionate is used for evaluation of the experimental data (Scheurer et al., 2005).

Human bodies

In order to test the applicability of the sheep model to human bodies, four selected human cases from the Institute of Forensic Medicine were examined

(Ith *et al.*, 2002; Scheurer *et al.*, 2005). The human cases were selected in order to match the ambient conditions of the animal model, i.e. were found in closed areas and the skull and brain were not injured. The forensic PMI was estimated by traditional forensic methods, i.e. by evaluating livor and rigor mortis, putrefaction signs and criminological information. After arriving at the Institute of Forensic Medicine the bodies were stored at 4°C for 20–70h before being examined by MRS. To make human data comparable to the sheep model, storage times in the cold have been subtracted from the total forensic PMI based on the experience that bacterial decomposition is massively reduced at low temperatures. The MR data acquisition was done the same way as for the sheep heads.

Figure 14.7 shows good qualitative agreement between the sheep model and human cases. The same metabolites occur at comparable points in time, in particular metabolites that are not found *in vivo* can be found postmortem (Ith *et al.*, 2001) in sheep as well as in human spectra. This is not self-evident since the bacterial colonization in an animal body could be different from humans. This qualitative agreement leads to the conclusion that the sheep model can be used to construct calibration curves, which then could be used for estimating PMIs in humans.

The four substances trimethylamine, propionate, butyrate and isobutyrate are generally not observed in healthy human brain tissue, however they are known as products of microbial activity. In addition, trimethylamine is almost exclusively found in bacterial metabolism (Brand and Galask, 1986). Succinate, another substance specifically seen in brain abscesses (Kim *et al.*, 1997; Sabatier *et al.*, 1999), could also be measured in sheep and human brain postmortem.

When calibration curves from sheep data are used to estimate PMIs in the four human cases (Figure 14.8), the estimated PMI can be compared with 'true' PMIs that have been mainly determined from criminal evidence. The comparison shows an acceptable agreement (Scheurer *et al.*, 2005), however the large error bars of forensic PMIs demonstrate that the determination of PMIs in forensic medicine is often very imprecise, covering time spans of several days or even weeks. This is an inherent problem and illustrates that a real 'gold standard' for validation of the MRS data is largely missing. It takes a prohibitively long time to wait for a large number of the rare cases where a human body is found after a long time, while it is still possible to restrict the time of death to a reasonable period. It will be necessary to collect these rare cases in order to validate the time curves established in the sheep model, however it is completely unrealistic to define the calibration curves directly in humans.

14.5   Outlook

The influence of environmental (i.e. external) factors such as ambient temperature, humidity, air motion, clothes, blankets etc. upon the processes of decreas-

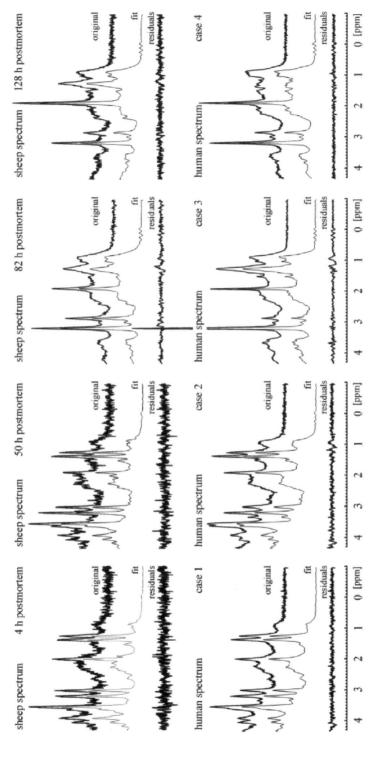

Figure 14.7 Spectra and fits of four human cases and corresponding spectra of the sheep model at comparable postmortem intervals, illustrating the similarity of the decay products. The residuals are quite flat, showing that the fitting procedure is quite accurate, however some remaining signals in the residuals indicate that there are still small amounts of undefined metabolites. Reproduced with permission from Scheurer et al. (2005)

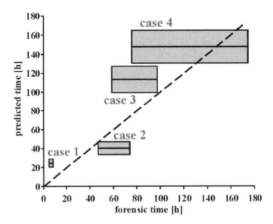

Figure 14.8 Comparison of predicted times determined by ^{1}H-MR spectroscopy versus times based on forensic evidence for the four human cases. Predicted PMIs are shown with an error range of ±2 standard deviations, while identity (predicted = forensic) is represented as a straight line. Reproduced with permission from Scheurer *et al.* (2005)

ing body temperature and decomposition is undisputed. However, in order to take these factors into account, it is necessary to consider additional information, e.g. from weather reports, etc. In forensic practice, the influence of temperature has been studied in detail (Friedrich, 1986; Althaus and Henssge, 1999; Al-Alousi *et al.*, 2001; Campobasso *et al.*, 2001), particularly the effect of temperature on specific processes, e.g. the degradation of muscle tissue, the increase of potassium in the vitreous humor, etc. Many authors also studied the effect of ventilation, humidity and clothing (Mann *et al.*, 1990; Campobasso *et al.*, 2001; Henssge 2002) and tried to classify the extent of the influence of these factors on the decomposition rate. Experience shows that the overall decomposition rate in tissues of human bodies stored at 4°C is significantly slower than at 20°C. However, since the changes observed by ^{1}H-MRS include two different mechanisms, i.e., autolysis and heterolysis, it is not at all obvious that the time courses of the metabolites from the two processes are affected in the same way by variations of the temperature. This is at the same time a complication as well as a chance. If the two processes proceed independently, the interpretation of the calibration curves would get much more complicated and, in turn, independent calibration curves for autolysis and heterolysis would contain additional information that could, in principle, be used to estimate the influence of the ambient temperature. In an ongoing study the influence of temperature on brain decomposition tissue is investigated with sheep heads in order to extend the sheep model.

Although it is generally accepted that internal factors may change the progress of decomposition, the influence of most of them is not well documented. Among them are antemortem diseases, which in some cases may even provoke death, such as diabetes or alcoholism. In contrast, the influences of drugs such as antibiotics have been proven to be insignificant for the later postmortem period since they are degraded within a short time (Daldrup *et al.*, 1982). The influence of other drugs is still unclear (Wagner, 1967). The cause of death can promote decomposition by facilitating bacterial colonization, i.e. in the form of injuries of the skin or enhanced blood fluidity in asphyxia. Systematic studies of the influence of the different factors are difficult, because they interfere and can hardly be standardized.

While it is obvious that calibration curves cannot be obtained with human cases, it is nevertheless necessary to validate the curves obtained from the sheep model with human cases at specific points of time. As described in the previous section, it is unlikely that the time of death can be determined exactly in a body that is only found several days or weeks after death. A delay in locating a human body is often due to limited or non-existent social contacts, which makes an exact determination of the PMI very difficult. Since restriction of the PMI is often based on criminal evidence that is not immediately available when the body is found, bodies have to be included in a study with the hope that a further restriction of the time span is possible. To date, 40 human bodies from the Institute of Forensic Medicine have been investigated by means of *in situ* ^{1}H-MRS of the brain – in many of these cases it is not possible to restrict the possible PMI to a short period in time, therefore it is necessary to acquire a larger number of bodies in future.

Despite the fact that many factors such as ambient temperature influence the results and that a gold standard is almost missing in human cases, the simple concept of a non-invasive and simultaneous observation of multiple metabolites during decomposition of the brain is promising. Since MRI is increasingly used to examine bodies in forensic investigations (Thali *et al.*, 2003; Yen *et al.*, 2004), the additional effort and expense to obtain a [1]H-MR spectrum of the brain are acceptable and can provide crucial information on the time of death, particularly during the later postmortem phase.

14.6   References

Al-Alousi, L.M., Anderson, R.A., Worster, D.M. and Land, D.V. (2001) Multiple-probe thermography for estimating the postmortem interval: I. Continuous monitoring and data analysis of brain, liver, rectal and environmental temperatures in 117 forensic cases. *J. Forensic Sci.* **46**: 317–322.

Althaus, L. and Henssge, C. (1999) Rectal temperature time of death nomogram: sudden change of ambient temperature. *Forensic Sci. Int.* **99**: 171–178.

Banaschak, S., Rzanny, R., Reichenbach, J.R., Kaiser, W.A. and Klein, A. (2005) Estimation of postmortem metabolic changes in porcine brain tissue using 1H-MR spectroscopy – preliminary results. *Int. J. Legal Med.* **119**: 77–79.

Benecke, M. (1996) [Expert insect identification in cases of decomposed bodies]. *Arch. Kriminol.* **198**: 99–109.

Berg, S. (1975) Leichenzersetzung und Leichenzerstörung. In: *Gerichtliche Medizin* (B. Mueller, ed.), Springer Verlag, Berlin, pp. 62–106.

Boesch, C. (1999) Molecular aspects of magnetic resonance imaging and spectroscopy. *Mol. Aspects Med.* **20**: 185–318.

Boesch, C. (2005) Magnetic resonance spectroscopy: basic principles. In: *Clinical Magnetic Resonance Imaging* (R.R. Edelman, J.R. Hesselink, M.B. Zlatkin and J.V. Cruess, eds), Saunders/Elsevier, Philadelphia, PA, pp. 459–492.

Bonte, W. (1978) *Der postmortale Proteinkatabolismus. Experimentelle Untersuchungen zum Problem der forensischen Leichenzeitbestimmung.* Habil. Schrift, Göttingen, pp. 57–75.

Bonte, W., Pohlig, K., Sprung, R. and Bleifuss, J. (1976) [The effect of microorganisms on protein catabolism in putrefaction studies]. *Beitr. Gerichtl. Med.* **34**: 173–178.

Brand, J.M. and Galask, R.P. (1986) Trimethylamine: the substance mainly responsible for the fishy odor often associated with bacterial vaginosis. *Obstet. Gynecol.* **68**: 682–685.

Campobasso, C.P., Di Vella, G. and Introna, F. (2001) Factors affecting decomposition and Diptera colonization. *Forensic Sci. Int.* **120**: 18–27.

Capron, A.M. (2001) Brain death – well settled yet still unresolved. *N. Engl. J. Med.* **344**: 1244–1246.

Chang, L., Ernst, T., Poland, R.E. and Jenden, D.J. (1996) In vivo proton magnetic resonance spectroscopy of the normal aging human brain. *Life Sci.* **58**: 2049–2056.

Coe, J.I. (1974) Postmortem chemistries on blood with particular reference to urea nitrogen, electrolytes, and bilirubin. *J. Forensic Sci.* **19**: 33–42.

Coe, J.I. (1989) Vitreous potassium as a measure of the postmortem interval: an historical review and critical evaluation. *Forensic Sci. Int.* **42**: 201–213.

Daldrup, T. (1983) [Practical experiences with the determination of cadaver age by evaluation of bacterial metabolic products]. *Z. Rechtsmed.* **90**: 19–25.

Daldrup, T. (1984) *Die Aminosäuren des Leichengehirns*, Enke Verlag Stuttgart, Stuttgart.

Daldrup, T., Hagedorn, H.J. and Korfers, M. (1982) [Microbiologic studies of fresh and putrefied cadaver brains]. *Beitr. Gerichtl. Med.* **40**: 379–382.

De Graaf, R.A. (1999) *In vivo NMR spectroscopy: Principles and Techniques*, John Wiley & Sons, Chichester, UK.

Diessner, H. and Lahl, R. (1969) [The post mortem determination of hydrogen-ion concentration in brain tissue homogenate and its relation to cause of death, course of death and time of death in selected autopsy material]. *Zentralbl. Allg. Pathol.* **112**: 162–170.

Doring, G. (1975) [Postmortem lipid metabolism]. *Beitr. Gerichtl. Med.* **33**: 76–84.

Endo, T., Hara, S., Kuriiwa, F. and Kano, S. (1990) Postmortem changes in the levels of monoamine metabolites in human cerebrospinal fluid. *Forensic Sci. Int.* **44**: 61–68.

Fechner, G., Koops, E. and Henssge, C. (1984) [Cessation of livor in defined pressure conditions]. *Z. Rechtsmed.* **93**: 283–287.

Forster, B., Ropohl, D., Prokop, O. and Riemer, K. (1974) Tierexperimente und an menschlichen Leichen gewonnene Daten zur Frage der Totenstarre. *Krim. Forens. Wissen.* **13**: 35–45.

Friedrich, G. (1986) Forensische postmortale Biochemie. In: *Praxis der Rechtsmedizin für Juristen und Mediziner* (B. Forster, ed.), Springer Verlag, Berlin, pp. 789–831.

Gadian, D.G. (1982) *Nuclear Magnetic Resonance and its Applications to Living Systems*, Clarendon Press, Oxford.

Glover, G.H. and Herfkens, R.J. (1998) The International Society for Magnetic Resonance in Medicine. Research directions in MR imaging. *Radiology* 207: 289–295.

Grassberger, M. and Reiter, C. (2001) Effect of temperature on Lucilia sericata (Diptera: Calliphoridae) development with special reference to the isomeg. *Forensic Sci. Int.* 120: 32–36.

Henssge, C. (2002) Todeszeitbestimmung an Leichen. *Rechtsmedizin* 112: 112–131.

Henssge, C., Althaus, L., Bolt, J., Freislederer, A., Haffner, H.T., Henssge, C., Hoppe, B. and Schneider, V. (2000a) Experiences with a compound method for estimating the time since death. I. Rectal temperature nomogram for time since death. *Int. J. Legal Med.* 113: 303–319.

Henssge, C., Althaus, L., Bolt, J., Freislederer, A., Haffner, H.T., Henssge, C., Hoppe, B. and Schneider, V. (2000b) Experiences with a compound method for estimating the time since death. II. Integration of non-temperature-based methods. *Int. J. Legal Med.* 113: 320–331.

Henssge, C. and Madea, B. (1988) *Methoden zur Bestimmung der Todeszeit an Leichen*, Schmidt-Römhild Verlag, Lübeck.

Howe, F., Maxwell, R., Saunders, D., Brown, M. and Griffiths, J. (1993) Proton spectroscopy *in vivo*. *Magnetic Resonance Quarterly* 9: 31–59.

Ith, M., Bigler, P., Scheurer, E., Kreis, R., Hofmann, L., Dirnhofer, R. and Boesch, C. (2002) Observation and identification of metabolites emerging during postmortem decomposition of brain tissue by means of *in situ* 1H-magnetic resonance spectroscopy. *Magn. Reson. Med.* 48: 915–920.

Ith, M., Kreis, R., Scheurer, E., Dirnhofer, R. and Boesch, C. (2001) Using 1H-MR Spectroscopy in Forensic Medicine to Estimate the Post-Mortem Interval: A Pilot Study in an Animal Model and its Application to Human Brain. *Proc. Int. Soc. Magn. Reson. Med.* 9: 388.

Kim, S.H., Chang, K.H., Song, I.C., Han, M.H., Kim, H.C., Kang, H.S. and Han, M.C. (1997) Brain abscess and brain tumor: discrimination with *in vivo* H-1 MR spectroscopy. *Radiology* 204: 239–245.

Kreis, R. (1997) Quantitative localized 1H MR spectroscopy for clinical use. *Prog. NMR Spectrosc.* 31: 155–195.

Krompecher, T., Bergerioux, C., Brandt-Casadevall, C. and Gujer, H.R. (1983) Experimental evaluation of rigor mortis. VI. Effect of various causes of death on the evolution of rigor mortis. *Forensic Sci. Int.* 22: 1–9.

Krompecher, T. and Fryc, O. (1979) [Determination of the time of death based on rigor mortis]. *Beitr. Gerichtl. Med.* 37: 285–289.

Kurthen, M., Linke, D.B. and Reuter, B.M. (1989) [Brain death, death of the cerebral cortex or personal death? On the current discussion of brain-oriented determination of death]. *Med. Klin.* 84: 483–487.

Lindlar, F. (1969) [Postmortem lipid changes and time of death determination]. *Beitr. Gerichtl. Med.* 26: 71–73.

Madea, B. and Henssge, C. (1991) Supravitalität. *Rechtsmedizin* 1: 117–129.

Madea, B., Kreuser, C. and Banaschak, S. (2001) Postmortem biochemical examination of synovial fluid–a preliminary study. *Forensic Sci. Int.* 118: 29–35.

Mallach, H.J. and Mittmeyer, H.J. (1971) [Rigor mortis and livores. Estimation of time of death by use of computerized data processing]. *Z. Rechtsmed.* **69**: 70–78.

Mann, R.W., Bass, W.M. and Meadows, L. (1990) Time since death and decomposition of the human body: variables and observations in case and experimental field studies. *J. Forensic Sci.* **35**: 103–111.

Marshall, T.K. and Hoare, F.D. (1962) Estimating the time of death: The rectal cooling after death and its mathematical expression. *J. Forensic Sci.* **7**: 56–81.

Marty, W. (1995) *Thermographie und Thermometrie in der Forensik mit besonderer Berücksichtigung der Todeszeitbestimmung,* Habil. Schrift, Zürich.

Mayer, M. and Neufeld, B. (1980) Post-mortem changes in skeletal muscle protease and creatine phosphokinase activity – a possible marker for determination of time of death. *Forensic Sci. Int.* **15**: 197–203.

Mittmeyer, H.J. (1980) [Muscle electrophoretic study in the determination of the time of death]. *Beitr. Gerichtl. Med.* **38**: 177–185.

Plattner, T., Scheurer, E. and Zollinger, U. (2002) The response of relatives to medico-legal investigations and forensic autopsy. *Am. J. Forensic Med. Pathol.* **23**: 345–348.

Provencher, S.W. (1993) Estimation of metabolite concentrations from localized *in vivo* proton NMR spectra. *Magn. Reson. Med.* **30**: 672–679.

Ross, B.D. and Danielsen, E.R. (1999) *Magnetic Resonance Spectroscopy Diagnosis of Neurological Diseases,* Marcel Dekker, New York.

Sabatier, J., Tremoulet, M., Ranjeva, J.P., Manelfe, C., Berry, I., Gilard, V. and Malet-Matino, M. (1999) Contribution of *in vivo* 1H spectroscopy to the diagnosis of deep-seated brain abscess. *J. Neurol. Neurosurg. Psychiatry* **66**: 120–121.

Scheurer, E., Ith, M., Dietrich, D., Kreis, R., Huesler, J., Dirnhofer, R. and Boesch, C. (2003) Statistical evaluation of 1H-MR spectra of the brain *in situ* for quantitative determination of postmortem intervals (PMI). *Proc. Int. Soc. Magn. Reson. Med.* **11**: 569.

Scheurer, E., Ith, M., Dietrich, D., Kreis, R., Husler, J., Dirnhofer, R. and Boesch, C. (2005) Statistical evaluation of time-dependent metabolite concentrations: estimation of post-mortem intervals based on *in situ* 1H-MRS of the brain. *NMR Biomed.* **18**: 163–172.

Schleyer, F. (1975) Leichenveränderungen, Todeszeitbestimmung im früh-postmortalen Intervall. In: *Gerichtliche Medizin,* (B. Mueller, ed.), Springer Verlag, Berlin, pp. 55–62.

Schneider, V. and Riese, R. (1980) Fäulnisveränderungen an Leichen. Ein Beitrag zur Todeszeitbestimmung. *Kriminalistik* **34**: 297–299.

Shapiro, H.A. (1965) The post-mortem temperature plateau. *J. Forensic Med.* **12**: 137–141.

Smith, K.G.V. (1986) *A Manual of Forensic Entomology,* British Museum, Natural History, London, and Cornell University Press, Ithaca, NY.

Thali, M.J., Yen, K., Schweitzer, W., Vock, P., Boesch, C., Ozdoba, C., Schroth, G., Ith, M., Sonnenschein, M., Doernhoefer, T., Scheurer, E., Plattner, T. and Dirnhofer, R. (2003) Virtopsy, a new imaging horizon in forensic pathology: virtual autopsy by postmortem multislice computed tomography (MSCT) and magnetic resonance imaging (MRI) – a feasibility study. *J. Forensic Sci.* **48**: 386–403.

Wagner, H.J. (1967) [On the influencing of time of death determinations and decomposition processes by drugs]. *Dtsch. Z. Ges. Gerichtl. Med.* **59**: 245–255.

Yen, K., Vock, P., Tiefenthaler, B., Ranner, G., Scheurer, E., Thali, M.J., Zwygart, K., Sonnenschein, M., Wiltgen, M. and Dirnhofer, R. (2004) Virtopsy: forensic traumatology of the subcutaneous fatty tissue; multislice computed tomography (MSCT) and magnetic resonance imaging (MRI) as diagnostic tools. *J. Forensic Sci.* **49**: 799–806.

Index

Molecular Forensics. Edited by Ralph Rapley and David Whitehouse
Copyright 2007 by John Wiley & Sons, Ltd.